I0462725

Rediscovering the Immune System as an Integrated Organ

by Peter Bretscher

 FriesenPress

Suite 300 - 990 Fort St
Victoria, BC, Canada, V8V 3K2
www.friesenpress.com

Copyright © 2016 by Peter Bretscher
First Edition — 2016

All rights reserved.

No part of this publication may be reproduced in any form, or by any means, electronic
or mechanical, including photocopying, recording, or any information browsing, storage,
or retrieval system, without permission in writing from FriesenPress.

ISBN
978-1-4602-7405-7 (Hardcover)
978-1-4602-7406-4 (Paperback)
978-1-4602-7407-1 (eBook)

1. Medical, Immunology

Distributed to the trade by The Ingram Book Company

Albert Einstein
Not everything that counts can be counted,
and not everything that can be counted counts

Preface

Immunology is one of the fastest growing of biological disciplines. We immunologists are gathering information at an unprecedented rate. Our contemporary ways of exploring the functioning of the immune system inevitably lead us to appreciate what appears to be the immune system's complexity. Current immunological papers are rarely accessible to outsiders, and their significance to the broad picture is often unclear even to specialists.

I have been doing research in the field of immunology for almost fifty years. The reason for writing this book is simple. I believe the complexity being revealed is not intrinsic to the subject, but rather a consequence of the culture within which contemporary immunological research is undertaken. The primary aim of this book is to make plausible an integrated and readily accessible, though necessarily incomplete, view of how the immune system functions. I hope this view will appeal to younger and perhaps even to some older immunologists, and members of the medical profession interested in an overview of the field. I also dare to hope some non-specialists will be intrigued and find this book both accessible and empowering.

The formulation of the Clonal Selection Theory in the 1950s and 1960s constituted an integration of observations and considerations at the level of the system with those at the cellular and molecular levels. The birth of molecular biology during these years was critical in bringing new molecular considerations to the fore. The formulation of the Clonal Selection Theory gave immunology a tremendous vitality over the next twenty years.

I argue, in contrast, that contemporary immunology is dominated by molecular and cellular considerations to such an extent that both older and newer observations and considerations, at the level of the system, are forgotten and ignored. Why has this scenario developed?

Immunological culture has changed dramatically in the last seventy years. The development of incisive techniques to examine phenomena at the molecular and cellular level has naturally transformed the field, with the potential for major advances. In many ways, the advance has been spectacular. This is evident to all who have experienced the progress of the last fifty to sixty years. However, I do not think this advance is occurring in an optimal manner. We are amassing enormous amounts of detailed information without gaining proportional insight. Why might this be?

It is obvious to most what is meant by a molecular and by a cellular concept or observation. This is because the entities of molecules and cells are relatively defined, usually with well-characterized properties, making descriptions of events in terms of cells and molecules concrete and compelling. These entities can be counted in the sense of Einstein's quotation. There are also observations, considerations, and concepts at the level of the system. The attribute of self-nonself discrimination—the ability of the immune system to respond more readily against foreign bodies than against self antigens—is a concept at the level of the system, and the existence of this attribute is supported by observations made at this level. Observations and considerations uniquely made at the level of the system often seem, to most contemporary immunologists, to be vague, in the sense of not concrete. They cannot be counted in the sense of Einstein's quotation.

I suggest this does not mean that observations and considerations at the level of the system are either imprecise or insignificant. I consider that Einstein's quotation, within the perspective developed here, can be translated to mean that *Some considerations/observations at the level of the system should be valued, and not all molecular/cellular observations are significant.* The currently popular perspective is illustrated by a comment recently made to me by a well-known immunologist: "You cannot, Peter, in this day and age, make a proposal that is not cast in molecular terms; it just cannot be taken seriously." I suggest the sustained adoption of such a mentality is leading to an ability to see only trees, and not the forests to which they belong.

It is my intention to demonstrate how diverse classical and contemporary observations and considerations at all three levels can be integrated to describe how the immune system functions in a coherent fashion. I explore and demonstrate how this integrated framework is also critical in devising strategies of medical intervention in the five areas of medicine that are immune system-related: allergies, autoimmunity, cancer, infectious diseases, and transplantation. I provide examples of successful strategies, whose aim is to prevent or treat some clinical states related to these five areas of medicine, that are based upon the integrated framework.

I have been driven to consider the contemporary sociological forces governing the field by a certain set of circumstances. These appear so surprising to me that I think it helpful to indicate their nature upfront. I think they illustrate the nature of major impediments to optimal progress.

A fundamental attribute of the immune system is that it has different means of fighting foreign invaders: primarily cell-mediated and antibody responses. These are referred to as different classes of immunity. Moreover, the immune system does not employ all classes of immunity at its disposal when fighting a foreign invader, but has a decision-making process to ascertain what class or classes of immunity to generate. A major question in immunology is the nature of this decision-making process. Moreover, it will be argued that an understanding of this process is central to the rational design of many medical interventions in the diverse fields of medicine related to the immune system.

A correct description of the decision-making process will allow us to understand why different variables of immunization affect the cell-mediated versus the antibody nature of the ensuing immune response. A number of these different variables were defined by immunizing live animals in studies carried out between forty to sixty years ago. These observations are observations made at the level of the system.

Three frameworks, cast at the cellular/molecular level, are currently accepted as plausible and employed to analyze the nature of this decision-making process. A description of one or more of these frameworks often constitutes the opening paragraph of papers in the highest-impact journals. However, I argue that these three current and popular frameworks, employed for over two decades, are inconsistent with the earlier observations made at the level of the system. It would appear either that the frameworks are wrong or that the forty to sixty year-old observations are incorrect. However, analogous observations have been made more recently and in ever more diverse systems. How can it be that this chronic state of affairs remains unresolved? In talking to colleagues, it seems that these older observations, and their potential generality, are no longer in the forefront of most immunologists' consciousness.

I believe it is the virtually exclusive focus on molecular/cellular observations and considerations, and the consequent employment of incorrect frameworks, that has led to such apparent complexity. *If incorrect frameworks are employed, the underlying coherence of the physiology of the immune system will be obscured.* The immune system will appear to be incredibly complex. I hope to convince the reader that, by paying attention to and incorporating observations and considerations at the level of the system, an integrated description of the immune system can be developed that is plausible and accessible to all. Given the likelihood that many contemporary immunologists will not be cognizant of what I consider to be critical and classical studies, I have decided to start with a clean slate. It never hurts to return to a re-examination of basic premises and to reconsider the significance of basic observations.

Acknowledgments

I started writing this book nine years ago. I would like to acknowledge the support and contribution of several people during different phases of the book's evolution.

Most important in the early stages were individuals who, though scientifically trained, were not professional immunologists. Thank you Iris Natyshak, Darrell Mousseau and William Albritton for your early enthusiasm and encouragement. Some immunologists were kind enough to read "middle versions" of the book. I am particularly grateful to David Nemazee for his very thoughtful comments. Colin Anderson is an immunologist who had/has different views from mine on some central questions. Colin gave me detailed scientific comments and this started a much-valued dialogue that has evolved into friendship. This dialogue has influenced the final text in several parts. I am indebted to both these colleagues. Lastly, two professional immunologists' response to very late drafts helped sustain me in the last mile. Zoltan Nagy is the author of a recent book entitled "A History of Modern Immunology: The Path Toward Understanding". I much enjoyed this book for both its science and its human values. I have never met Zoltan personally, but we have conversed by email subsequent to my reading his book. On reading a late draft of mine, Zoltan strongly encouraged me to get my unique message to the immunological community. Ted Steele, who I had known as an enthusiastic young postdoc in the mid 1970s, recently got in touch with me. I gave him a version very close to the final one. His enthusiasm and comments made the last stages of preparing the manuscript a time of anticipation rather than one of drudgery.

The enormous undertaking of writing a book such as this would be insurmountable without help. I thank Juliane Deubner, Medical Illustrator, Saskatoon, for doing such a nice job on the Figures, and showing patience with me. I also thank the professional help of diverse individuals from FriesenPress.

This book reflects more of my life than the last nine years. Two individuals critically helped in our research over extended periods of time and also, by the nature of their personality, made every morning a pleasure. Thank you Mohammed Dhalla and Guojian Wei. You both know what a central role you played in our investigations at different times.

I wish to acknowledge another source of inspiration. I thank my undergraduate and graduate students. Teaching is a passion of mine. I have taught an undergraduate course on the Principle's of Immunology over the last thirty years. I always hope my sessions, as I call my "lectures", will empower my students to realize that nature is open to tentative understanding, and that this experience will empower them. I use the arts accumulated over the years to achieve a dialogue with my undergraduate students, in contrast to "teaching" them. The story I tell in this book illustrates that the same type of dialogue occurred with my graduate students. These interactions with my students have taught me much about how to make ideas intelligible and clear both to others and to myself. I am indebted to those students who asked me questions, and who thus inspired me as I perceived their intellectual awakening.

Lastly, I want to thank our son Paul. Paul's training is in the social sciences. I am fascinated by the sociology of science. Paul has only read a few pages of my drafts, such as the Preface. My conversations with him over the years, about how progress is perceived in the political sciences, influenced me. It gave me strength to see the evolution of immunology from a more sociological point of view. These discussions gave me confidence to develop and elaborate upon intuitive ideas as to how social forces affect immunological progress.

Table of Contents

A note on technical terms

My teaching experience has led me to try to develop ways of minimizing the barriers that technical terms present in achieving a valid understanding. This book has a glossary at the end, but it does not provide definitions of terms used, but rather the page(s) in which the term is defined and its use shown in context. The *term*, when first used and its meaning explained, is ***bolded and italicized.*** I believe it is critical to use terms as precisely as possible to foster clear discussion and understanding. However, precision is only possible in a context, and so definitions can become somewhat elaborate and, in the end, circular and so not absolute. I hope this way of defining terms in context will facilitate the reader's understanding.

Chapter 1

AN INTRODUCTION TO THE IMMUNE SYSTEM

A context for understanding some salient characteristics of the immune system and their pertinence to medicine

I attempt in this book to develop an overall picture of how the immune system functions. I begin by outlining two pivotal questions as to how immune responses are regulated. I explain how these questions arose, and justify their centrality. I argue that the answers to them are necessary to provide a rational basis for diverse advances in medical prevention and treatment. I shall consider their pertinence to the five fields of medicine that are related to the immune system. This is not a gentle introduction, but rather one that attempts to make plausible, at the outset, the importance of two central questions for achieving effective medical interventions.

Paul Ehrlich and Louis Pasteur, pre-eminent among the founders of immunology, made their contributions from the mid-1800s to the early 1900s. Ehrlich and his colleagues reported in the early 1900s on their findings that antibodies could be generated upon immunizing one goat with red blood cells from another goat. These antibodies reacted with, and could destroy, the donor goat's red blood cells, but antibodies against the red blood cells of the immunized goat could not be detected.[1] Ehrlich, reflecting on these findings, came to the thought that, if such antibody were produced, the consequences would be devastating. He imagined that there must be processes to ensure that such antibody, specific for the host's own red blood cells, is not produced. He recognized that the immune system had this ability to respond against foreign antigens, but not against self antigens that are part of the body to which the immune system belongs.[1] We refer to this attribute of the immune system as self-nonself discrimination.

Ehrlich's insight was first fully appreciated some decades later, beginning in the 1940s, when it became recognized that some people are ill because the immune system attacks cells or molecules of the body to which the immune system belongs.[2] One such disease is **autoimmune hemolytic anemia**, in which antibodies to the patient's own red blood cells are produced, resulting in their destruction and hence the anemia.[2] What could be the basis of the physiological attribute of self-nonself discrimination? Considerations at the level of the system are enlightening.

We know that organ grafts from parents or siblings are liable to be rejected. On reflection, this is remarkable. Consider the particular case of a kidney grafted from one brother to another. The grafted kidney is different from the recipient's own in virtually all cases, as the brothers are genetically distinct. The immune system recognizes differences between the donated and the recipient's own kidneys and rejects the donated graft. However, in the luck of the draw that occurs at conception, the genes that make the brother's kidney different from the recipient's own kidneys could have been handed down to the recipient. This line of consideration allows us to realize that the immune system has the *intrinsic ability* to reject an individual's own kidney, or any other organ.

However, this does not normally occur. There must therefore be a mechanism by which the immune system learns that an individual's own kidneys are "self", and that the brother's kidney is foreign or "nonself". This mechanism would account for the attribute of **self-nonself discrimination,** recognized by Ehrlich in the early 1900s. Foreign molecules, such as those associated with a brother's kidney, are generically called foreign antigens, whereas the corresponding molecules of the body's own kidneys are referred to as **self antigens**. The presence of **foreign antigens** provokes an immune response, whereas the presence of self antigens prevents such responses. *The first of the two questions I consider pivotal is this: how do self and foreign antigens interact differently with the immune system, resulting respectively in the prevention and provocation of immune responses?*

The second question can be appreciated in a certain context. Some individuals become allergic to house dust mites because they make a particular kind of immune response to these foreign antigens; others, living in the same geographical area and so exposed to the same house dust mite antigens, do not become allergic. Why? It turns out that both these allergic and non-allergic individuals make antibody responses against the house dust mite antigens. As we shall see, the class or type of antibody made in **allergic** and **non-allergic** individuals have different properties and functions. This situation reflects a more general attribute of the immune system.

Observations in diverse situations show that the immune system has a few major ways of fighting foreign invaders. Moreover, the immune system does not, when it attacks a foreign invader, employ all the means at its disposal, but has a decision-making process to decide which to use. Immunologists thus say that there are distinct classes of immunity that are differentially regulated. Therefore, under some circumstances there may be a predominant cell-mediated response, referred to as **cell-mediated immunity,** and under other circumstances a predominant humoral response. This is a response mediated by antibodies and often referred to as **humoral immunity**. In the case of individuals allergic and non-allergic to an antigen, and living in the same geographical area and so exposed to the same antigens, the class of antibody produced is different.[3]

The second question I consider pivotal is: what is the nature of these decision-making processes that determine the class of immunity generated?

We now consider why the answers to these two questions may provide a basis for the prevention or treatment of situations of medical interest.

The attribute of self-nonself discrimination sometimes fails. In this case, the immune system generates immunity against self antigens, a phenomenon known as autoreactivity. If the **autoreactivity** results in damage to the host, it is referred to as **autoimmunity**. One example of autoimmunity, already described, is autoimmune hemolytic anemia. Another well known example is **autoimmune diabetes**, a disease in which the immune system fights the β-islet cells of the pancreas that produce insulin, the hormone that controls glucose metabolism.[4] A rational understanding of how to prevent or treat autoimmunity requires an understanding of how antigens naturally provoke or prevent immune responses, i.e. to understand the basis of immunological self-nonself discrimination. Similarly, such an understanding should allow one in principle to overcome the barrier to transplantation of foreign organs.

The three other areas of medicine, besides autoimmunity and transplantation, with a close relationship to the immune system are allergies, infectious diseases, and cancer. I argue that in most clinical situations related to these three areas, successful intervention will be greatly helped by understanding the basis of immune class regulation.

Consider first infectious diseases. Immunologists understand why vaccination against certain pathogens, such as those causing smallpox, polio, diphtheria, and tetanus, is effective. We know that antibody that recognizes the pathogen or toxic products that the pathogen produces is effective in preventing the corresponding disease. Vaccination generates a "memory imprint" upon the immune system, often resulting in the long-term presence of antibody and in a greater and more rapid antibody response upon natural infection. This antibody response curtails the spread of the pathogen and facilitates the neutralization and removal of toxic products produced by the pathogen. Consequently, the pathogen and its products are kept at bay, and so disease does not develop.

However, this pattern is not observed in all infections. Clinical observation shows that those infected by the pathogen responsible for **acquired immune deficiency syndrome (AIDS)**, i.e. the **human immunodeficiency virus (HIV)**, do not invariably become ill. Those who become ill upon infection are called **patients**, and those who do not are called **healthy infected**. It turns out that most healthy infected individuals make a predominant, sustained, and stable cell-mediated response,[5] whereas patients eventually make a mixed cell-mediated/antibody or a predominant antibody response. As discussed later, similar correlations occur in other chronic diseases caused by several other significant intracellular pathogens.[6] I argue that, if vaccination is to be effective in these cases, it must cause an imprint upon the immune system that guarantees a stable and predominant cell-mediated response upon natural infection. I also argue that our understanding of how the class of immunity is determined has allowed this to be achieved in a number of instances, and to demonstrate the efficacy of such **cell-mediated imprints** in protecting against disease that would otherwise follow upon natural infection.

Studies in the last three decades have provided unequivocal evidence that the immune systems of cancer patients respond against their cancers,[7,8] as envisaged by Paul Ehrlich in the early 1900s.[9] We shall consider the hypothesis that cancer can most often be effectively contained by a strong and predominant cell-mediated response, whereas cancer progression is often associated with either too weak a cell-mediated response, or with an immune response that has a substantial antibody component. This hypothesis accounts for many observations made in the field of cancer immunology.[10] We shall argue that insight into what is required to generate immune responses in general, and into the basis of immune class regulation, are central to the development of rational strategies to both prevent and optimally treat cancer.

Lastly, we have seen that individuals living in the same geographical area, some allergic and others non-allergic to house dust mite antigens, produce different classes of antibody. It again seems likely that an understanding of how different classes of antibody are generated will allow one to ensure that the immune systems of individuals, genetically predisposed to become allergic to house dust mite, can be **imprinted** to produce, upon natural exposure, the classes of antibody produced by non-allergic individuals. Similarly, treatment of an allergy may be achieved by modulating the class of antibody produced by allergic individuals to that class produced by non-allergic individuals.

The evolutionary context in which the immune system arose

The adaptive immune system is first seen in vertebrates. Non-vertebrates also have diverse defense mechanisms, referred to as mechanisms of innate defense. Vertebrates have both immune and innate mechanisms of defense. The pre-existence of these innate mechanisms, as the immune system evolved, is recognized as having profoundly affected the nature of the immune system in at least two respects.

First, it takes several days, if not weeks, for an effective immune response to be generated against a foreign invader. Bacteria and viruses are common invaders. A bacterium typically multiplies in less than an hour under favorable conditions. At such a rate, one bacterium would give rise to about 10^{30} bacteria in five days, weighing about 10^{18} kilograms. Nothing like this normally happens, of course, because innate mechanisms of defense contain the bacteria, or at least greatly limit their expansion, and because favorable conditions for bacterial replication would not be sustained over this time.

The debilitating consequences of genetic defects in vertebrates that compromise innate defense mechanisms attest to their supreme importance.[11] Without mechanisms of innate defense, the immune system as we know it would not exist, nor would we. Thus, the leisurely tempo of immune responses is only possible because innate defense mechanisms initially hold the fort against most foreign invaders with considerable efficacy.

Second, the immune system adopts and adapts many of the mechanisms of innate defense in providing protection. We shall explore two examples of such adoption in the next section.

The five attributes of the immune system[12]

As already mentioned, vertebrates and non-vertebrates share a variety of innate defense mechanisms, but immune systems can be distinguished from the systems of innate defense by five attributes. Moreover, we shall see that most of these five attributes are interdependent; it is difficult to envisage how one could have arisen independently of most of the others. However, to set the scene for such considerations, I shall first describe two examples of innate defense mechanisms, two of the several that are central to the realization of protection provided by the immune system. I use these examples to illustrate general features of innate defense mechanisms, and how they differ from those of the immune system. We shall also see in more detail how the immune system adopts and so exploits these mechanisms.

The simple multi-cellular organism known as a hydra, which predominantly consists of a bilayer of cells, one opening to the outside operating as its mouth and anus, has mobile cells whose function is to ingest and destroy foreign invaders. These cells use harsh conditions and enzymes to break down the invaders into small molecules and thereby to kill them. These cells thus protect hydra against invaders. The small molecules generated from digesting the invaders are used as building blocks to synthesize the small and large molecules the cells need to survive and multiply. Similar cells occur in all multi-cellular forms of life. These defensive cells are called

phagocytes ("eating cell" from the Greek "phago", I eat), and the process of ingestion and digestion of foreign matter by cells is called *phagocytosis*. Vertebrates have
several types of phagocytes in their bodies, and these cells constitute a major barrier
to the establishment of infections. Phagocytosis is thus a phylogenetically old and
important mechanism of innate defense. A first step in the phagocytic process is to
envelop the invader, a bacterium for example, and to take the enveloped invader into
the phagocyte. This first step is referred to as *opsonization.*

Gardeners will have observed how a substantial scratch on their skin results in
swelling within a minute or two. The skin provides a protective physical/chemical
barrier against foreign invaders. A gardener's scratch provides microorganisms with
a splendid opportunity to break through this barrier. However, a series of events is
triggered by the pressure of the sharp object on our skin that leads to the scratch.
First, the scratch becomes red within a minute and swelling or edema can be seen
along its length within minutes. The pressure on the skin leads *pressure nerve
cells*, nerve cells that respond to pressure, to locally release a substance that acts
on *mast cells*, strategically placed close to the nerve, to cause the rapid release of
substances that cause the local blood capillaries to dilate and become leaky. This
allows defensive cells such as phagocytes and fluid containing diverse anti-microbial
molecules to accumulate around the site of injury, leading to the observed redness
and swelling, technically referred to as *inflammation*. These phagocytic cells and
anti-microbial molecules provide a second line of defense after the first, the skin, has
been breached. This whole process occurs quickly, within minutes, and is therefore
called *acute inflammation.* We shall later see how phagocytosis and acute inflammation, as well as other innate defense mechanisms, are critical to the working of the
immune system.

1. Positive memory as an expression of the attribute of adaptability

These two examples of innate defense, phagocytosis and acute inflammation,
exemplify some characteristics that distinguish these mechanisms from those of the
immune system.

First, these and other mechanisms of innate defense act quickly following the
triggering event. Acute inflammation is evident within minutes following the formation of a scratch, and the engulfment of a bacterium, following the binding of the
bacterium to the surface of a phagocyte, takes minutes. The protection provided
is said to be *constitutive* or immediate, as all the elements of the mechanism are
present and ready upon the first encounter with an insult, or within minutes thereafter. This constitutive quality is in contrast to the defense provided by immunity.
It takes several days if not weeks for the immune system to develop the means to
provide efficient protection following an infection.

The existence of this *inductive* period obviously reflects processes required to achieve immunity. We refer collectively to these processes as the ***induction of an immune response***. This phenomenon reflects the ***adaptability*** of the immune system. The term adaptability is operationally used in immunology to refer to situations where immunity against an antigen depends upon the history of an animal or person with respect to this particular antigen. Thus, exposure two weeks previously to a pathogen—the recent history of the individual—can result in effective immunity to a pathogen that is not evident on the day of infection.

The Greek historian Thucydides recorded observations in about 500 BCE that also reflect the adaptability of the immune system. He noted that those who had had the plague, often many years previously, could tend the sick with impunity. He recognized that individuals who had been infected but survived an epidemic with a high mortality rate were resistant to a subsequent wave of the same epidemic, and were immune. These observations illustrate the immune system's attribute of ***positive memory***, in that a ***secondary response*** to an antigen upon a ***secondary infection*** is usually more rapid and of greater intensity than the ***primary response*** that occurs on a first or ***primary infection***.

This is another and most important example of the attribute of adaptability, different from the first, described above. This memory does not represent a state of immunity, but rather an ability to generate more intense immunity at a more rapid rate than the immunity that occurs upon a primary infection. This memory generally lasts for years. It is the primary basis of successful vaccination against, for example, small pox and polio. What other body systems show such striking adaptability? The most obvious example is our brain, the higher nervous system. Someone who has learned French will respond differently to a jest in that language than one who has not. We shall also see that there are other forms of adaptability by the immune system that are critical to its functioning.

2. The attribute of specificity: the discovery of antibody and how it led to an appreciation of this attribute

The development of vaccination

Voltaire reported from his travels in the early- to mid-1700s on the folk practices, current in China and in the Middle East, that were able to protect against the ravages of smallpox. We now know this disease is caused by the smallpox or variola virus.[12] Material harvested from the crusts of the skin-pocks of patients, when administered to an uninfected individual, protects this "immunized individual" from smallpox. It is estimated that this "vaccination" process had a mortality rate of about 1%. This might seem unduly hazardous to the modern mind, but it must be judged in the context of the devastating effects of the disease itself. This was a time when an estimated sixty of every hundred individuals in England contracted smallpox and, of this sixty, twenty died and twenty were permanently disfigured by pock-scars consequent to infection.[12]

Cows can suffer from a similar disease, cowpox, caused by a virus similar to human smallpox virus, but the cowpox virus, vaccinia, is not virulent in humans. When humans are infected with vaccinia virus, they do not become seriously ill. It was commonly recognized in the 1700s that individuals who had naturally come into contact with cowpox were resistant to smallpox. This recognition is reflected in a refrain of a ditty of the time: "Where are you going, my pretty maid? I'm going a-milking, Sir, she said." Edward Jenner undertook a systematic investigation to determine whether immunization with the cowpox virus, obtained from the pustules of infected cows, could protect humans against smallpox. He showed in the late 1700s that such immunization did indeed provide protection.[12]

About fifty years later, in the mid- to late-1800s, Robert Koch and Louis Pasteur demonstrated that many diseases, some leading to death, are due to infectious organisms. This demonstration required Koch and Pasteur to develop the means of culturing infectious organisms. The cultured organism would then cause the pertinent disease when injected into an appropriate animal, leading the animal to suffer a series of well-defined and characteristic symptoms, sometimes resulting in morbidity.

Organisms able to cause severe disease are said to be **virulent**. Pasteur found a variety of conditions, often non-optimal for growth, under which he could propagate these infectious organisms. Sustained culture under such unfavorable conditions often led to the organism's loss of **virulence**, so that the full symptoms of the disease were no longer obtained upon injecting these organisms into an animal. The infectious organisms are said to be **attenuated**. Pasteur employed such attenuated microorganisms to protect against the corresponding virulent organisms, just as Jenner had protected against the ravages of the smallpox virus by immunization with the harmless cowpox or vaccinia virus. Pasteur called his process of immunization by infection with attenuated pathogens **vaccination,** in honor of Jenner's studies with vaccinia/cowpox virus, (vacca being Latin for cow).[12]

These discoveries represent the birth of immunology and of the study of infectious diseases. Progress was now rapid in many areas. Pasteur proposed in the late 1800s a concise idea for how vaccination works. His idea was that the body of uninfected people contains vital foodstuffs required for an attenuated or a virulent organism to thrive or even to survive. Pre-exposure to an attenuated microorganism would deplete or exhaust the vital foodstuffs, and so a subsequent infection with a virulent organism would not now allow the organism to thrive and cause disease.

Pasteur's **Exhaustion Hypothesis** was soon found wanting. Some pathogenic bacteria, such as diphtheria bacilli, exert their pathogenicity by producing **toxins**. Some of these toxins accumulate in the culture fluid in which the bacteria are grown. Thus the symptoms of the disease can be produced either by infecting animals with the virulent bacteria, or by administering the bacterial toxin. Immunizing with **toxoid,** a slightly altered, non-toxic form of the toxin, protects animals against a subsequent challenge of the toxin that is lethal in unimmunized animals. This finding was at odds with Pasteur's Exhaustion Hypothesis, as a non-living molecule was hardly likely to deplete the vital foodstuffs. Moreover, immunization with dead bacteria could also lead to protection.[12]

Antibodies: the mediators of protection

The whole field was transformed when it was found that this immune resistance to a toxin, or to a bacterial challenge, could be transferred from an immune, and hence resistant, animal to a non-immunized animal by giving serum obtained from the immunized donor to the naïve recipient. I shall spell out in detail what is meant by this crucial statement. Mouse 1 of Figure 1 is immunized with toxoid, making the mouse resistant to a normally lethal challenge of toxin. This resistant mouse is then bled, the blood allowed to clot, and the honey-colored, cell-free serum harvested, leaving behind the congealed, red mass of clotted cells. This immune serum is administered intravenously to an unimmunized recipient mouse, see mouse 3 of Figure 1. Mouse 3 now resists a challenge of toxin that is lethal when given to a normal, unimmunized mouse, mouse 2 of the figure.

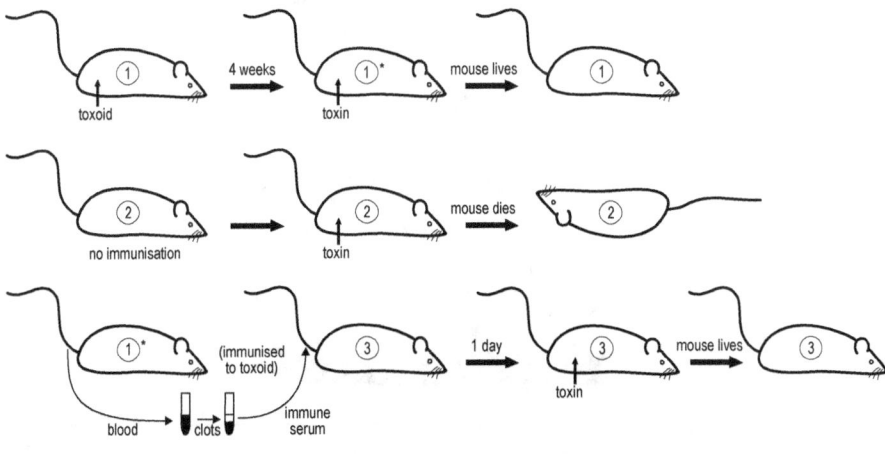

Fig. 01. Illustration of how passive immunity is achieved.

Animals could thus be made resistant in one of two ways. They could be immunized with toxoid or a non-pathogenic challenge of the bacteria, in which case their immune system would be responsible for producing whatever it takes to provide protection. This process is called **active immunization**, as it requires the immune system to be active. The resistance to a toxin, conferred upon an animal as a result of receiving serum from an immune and resistant animal, is called **passive immunization,** as the acquisition of this immunity does not require the immune system of the recipient to be active. The existence of passive immunity led people to envisage that there are protective molecules in immune serum. Various experiments soon showed this inference to be correct; such protective molecules were called **antibodies.**

The recognition that immune serum contains molecules that can protect against pathogenic bacteria naturally led to an examination of what happens when such serum is added to a suspension of the bacteria. Conditions were found, namely incubation at our body temperature of 37^0C, under which the bacteria disappear within minutes. This disappearance is manifest upon microscopic examination, and by the subsequent inability to grow bacteria from the suspension. No wonder immune serum is protective! We now know that the antibodies present in blood and in serum bind to the bacteria, allowing another series of molecules found in blood and serum to bind to those antibodies in a series of piggy-back reactions. This collection of molecules is called **complement.** The complement components, brought close to the bacterial surface in this reaction, act to punch holes through the bacterial membrane, see Figure 2. This subsequently leads to the breaking apart of the bacteria or to **bacterial lysis.**

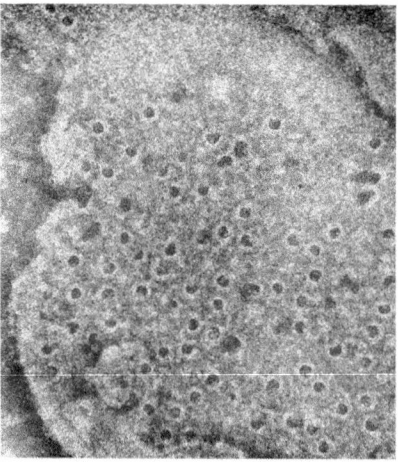

Fig. 02. Electron micrograph of a bacterial membrane with holes caused by antibody-dependent complement-mediated attack. From reference 12

Antibody-dependent, complement-mediated cell lysis is another immunological means of fighting bacteria, in addition to phagocytosis already described. It is interesting to note that complement also has several functions in innate defense that predate the evolutionary appearance of the immune system. Complement from non-vertebrates and vertebrates can bind directly to many bacteria to cause their lysis, though relatively inefficiently.

The discovery of passive immunity was seminal. It led to the recognition of antibodies as the mediators of protection and so, in the long run, to their characterization. Most important was the discovery that honey-colored, transparent, and liquid serum obtained from an animal immune to a toxin would go cloudy when added to a transparent solution of the toxin, and a white precipitate would form. This process is called the *precipitin reaction*. It is a way of detecting antibodies specific for soluble molecules to which the antibodies bind. Anything that can be recognized by antibodies is called an *antigen*. Thus, toxins can be recognized by antibodies and are said to be *antigenic*. A molecule able to induce the production of antibodies upon injection into an animal is said to be *immunogenic*—capable of generating an immune response. Toxins in general are both antigenic and immunogenic.

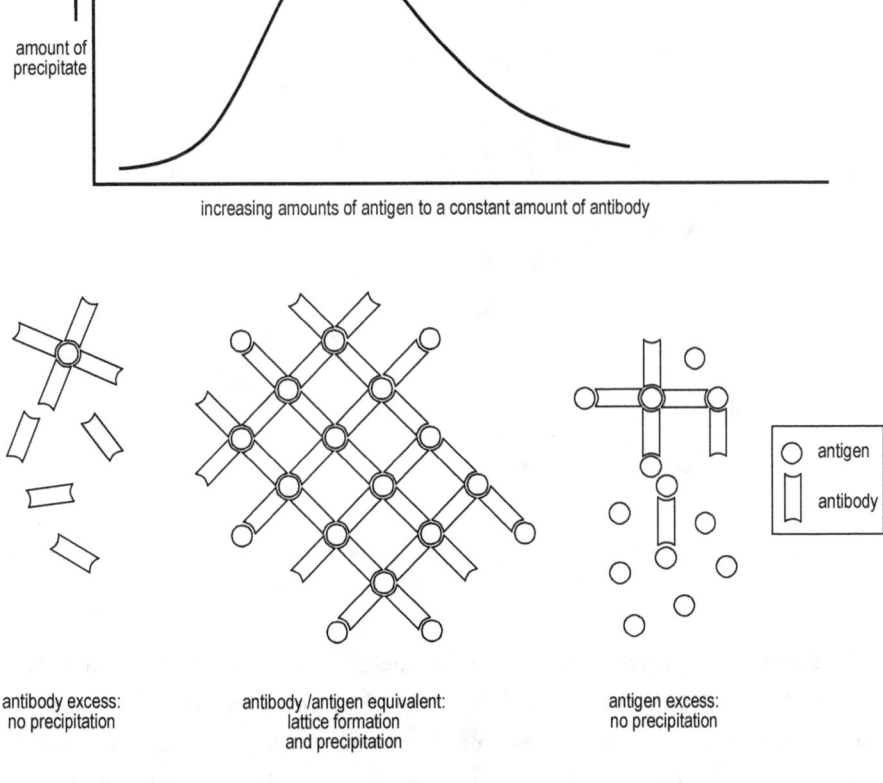

Fig. 03. How Marrack's Lattice Hypothesis accounts for the precipitin reaction between antigens and antibodies

Interestingly, it was as long as sixty years after the discovery of the precipitin reaction that Marrack came to a correct realization of its basis. The basic unit out of which all antibody molecules are constructed has two identical halves. Some antibodies consist of one such unit, others of several. All antibody molecules can thus bind at least two antigen molecules.

Moreover, antigens are usually fairly large molecules, and can themselves bind to more than one antibody molecule, usually several. This means that antibodies and their antigens can form big nets or lattices in which the antigen and antibody molecules are linked together to form large structures, as illustrated in Figure 3. Marrack envisaged that such large structures would be insoluble and precipitate, thus accounting for the precipitin reaction. Many lines of evidence provide compelling support for this **Lattice Hypothesis**, including the direct visualization of lattices formed when antibody and antigen are mixed together and the resulting structures examined by electron microscopy[13], see Figure 4.

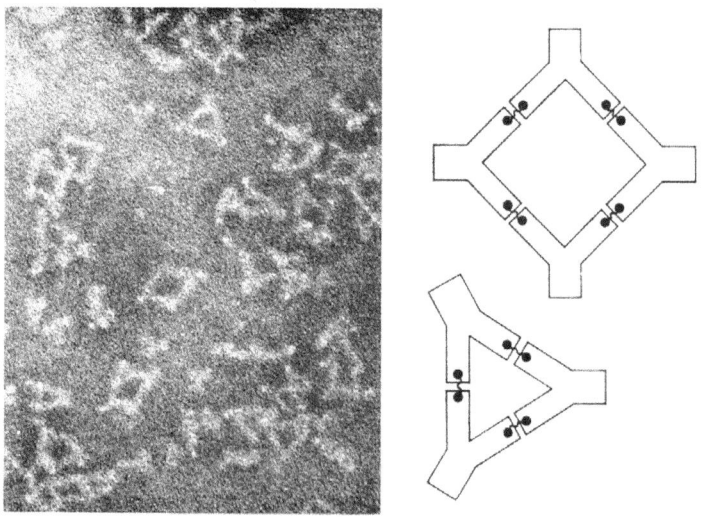

Fig. 04. Electron microscopic image of small lattices of antibodies with a very small, divalent antigen and an interpretation of the images observed. Adapted from reference 12

The antigen employed in this study was specially synthesized to have only two sites to which antibody can bind, and this artificial antigen is much smaller than most natural ones. In these circumstances, small lattices tend to form, which can be readily visualized with an electron microscope. In practice, in order to obtain large lattices with a multivalent antigen and to generate a visible precipitate, equivalent amounts of antibody and antigen must be present. This is shown by an experiment in which several tubes are set up, into each of which is put the same amount of antibody. A small amount of antigen is added to tube 1, and increasing amounts are added to tubes labelled with a larger number. The top part of Figure 3 shows how the amount of precipitate formed depends upon the amount of antigen present. The lower part illustrates the explanation of this dependence in terms of the lattice hypothesis.

The exquisite specificity of antibodies

It became apparent in the early 1900s that the way the immune system recognizes foreign entities is radically different from anything previously imagined. The immune system has an incredible ability to distinguish subtle differences between the structures on various cells and between closly related molecules. In this sense, the immune system is more specific than innate mechanisms of defense. An understanding of how the exquisite specificity of antibodies came to be appreciated is central to an understanding of one of the most remarkable features of the immune system. We will now trace how an appreciation of antibody specificity came to be recognized.

It was realized by the end of the 1800s that antibodies are specific. Thus, antibodies raised against diphtheria toxin did not react with tetanus toxin, as measured by the precipitin reaction, and vice versa. However, there had to be a limit to specificity. For example, immunization with the cowpox virus provides protection against the smallpox virus and so, as protection is due to antibodies, some of those generated upon immunization with the cowpox virus must be able to bind to the smallpox virus. When antibodies can bind to two similar antigens, the antigens are said to **crossreact**. The antibodies that recognize the common structures are called **crossreactive antibodies.**

Next to nothing was known about the structure of such large molecules as the bacterial toxins in the late 1800s and early 1900s. However, chemists such as Kekule had recently elucidated the structure of some simple organic molecules. Given both these limitations and this new knowledge, people tried to gain insight into the nature of the interaction between antibodies and antigens. They took small organic molecules of known structure and with an approximate size of a benzene ring or two in a reactive form, so they would form a covalent bond with a protein molecule that was itself **immunogenic**, i.e. able to induce an antibody response. They could then raise antibodies to the modified, conjugated antigen. Did this result in antibodies being generated that were able to recognize the small molecule itself?

An ingenious approach showed such antibodies are generated. Two large immunogenic proteins were chosen, say A and B, that did not crossreact; thus anti-A antibodies would precipitate with A but not with B, and anti-B antibodies would precipitate with B but not with A. The protein A was conjugated with a chemically reactive form of a small molecule, generically called a **hapten**, and denoted as "h", to form the h_n-A conjugate, where n represents the average number of small molecules attached to one molecule of A.

This conjugate was then employed to immunize rabbits. An antibody response was induced, as shown by the fact that the serum from immunized but not unimmunized rabbits would precipitate with both A and with h_n-A. It was also found that the immune serum did not precipitate with the antigen B, as expected, as A and B were chosen not to crossreact, but it did precipitate when added to the conjugate h_m-B, which contained on average m molecules of h for every molecule of B. This seemed to suggest that there were antibodies able to recognize the small hapten h. The surmise that the precipitation of anti-(h_n-A) with h_m-B was due to anti-h antibodies was later confirmed in a stunning manner. The precipitation of the anti-(h_n-A) antibody with h_m-B could be inhibited by a sufficient presence of the hapten h in a chemically non-reactive form. Moreover, this inhibition was very specific. Figure 5 depicts how the lattice theory of the precipitin reaction explains this inhibition. Briefly, the hapten, small in size and able to bind to only one binding site of an antibody molecule, is **monovalent**. The hapten therefore cannot interact with anti-hapten antibody to form lattices. However, the hapten can bind to anti-hapten antibody, blocking its binding sites. Excessive amounts of hapten can thus prevent the anti-hapten antibody from interacting with h_m-B to form lattices with this antigen.

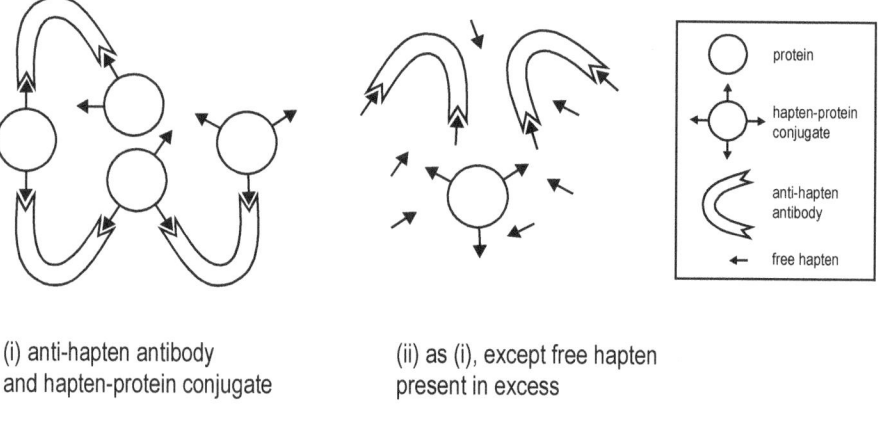

(i) anti-hapten antibody
and hapten-protein conjugate

(ii) as (i), except free hapten
present in excess

Fig. 05. Illustration of how a hapten can inhibit the precipitation of anti-hapten antibody with a multi-haptenated protein conjugate

In the early 1900s, Landsteiner employed haptenated antigens to raise anti-hapten antibodies and to examine their specificity.[14] He showed that molecules structurally only very slightly different from h, denoted as h′, were often unable to precipitate anti-h antibodies, as assessed with h′-B conjugates, whereas h itself, in the form of h-B, was able to do so. Figure 6 summarizes some of his observations. It is apparent how extraordinarily specific these antibodies can be. These observations led to profound realizations. It is worth pausing to consider their significance.

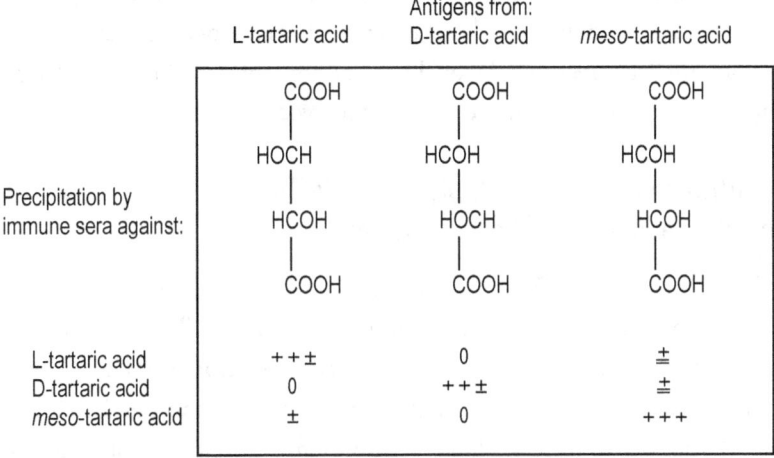

Fig. 06. A demonstration of the specificity of antibodies after Landsteiner. +++ and ++± represents strong precipitation, 0 represents no precipitation and ± and +/= partial and very partial precipitation. Taken from reference 12

3. The Attribute of Universality

These studies, and the resulting appreciation of the exquisite specificity of antibody molecules, led to two different but related ideas. The first was that the immune system could produce antibody against any small foreign molecule if an animal was appropriately immunized. Antibody could be raised against an incredible diversity of small molecules, or haptens, if they were chemically coupled to an *immunogenic carrier*, and if this *hapten-carrier conjugate* was employed to immunize, say, a rabbit. These small molecules could bind antibody and therefore were said to be antigenic but, when administered directly to an animal, did not lead to the production of antibody and were therefore said to be *nonimmunogenic.* These small haptens included new molecules recently invented by organic chemists. The immune system seemed able to respond to the unknown and the unanticipated.

The ability to respond to virtually every foreign molecule bigger than a certain size is amazing, and is referred to as the attribute of *universality*. An everyday analogy of universality would be a large clothing store that is able to provide well fitting suits for people of all shapes and sizes. How can universality be achieved? We will consider this later.

The recognition of the universality of the immune response, and of the great specificity of antibodies, led to the second realization: there must be an astronomical number of different antibody molecules. We now know that the number is in the order of hundreds of billions. A partial recognition of the enormity of this number naturally led to the question of how such a *diversity* of antibodies could be *generated.*

One of the triumphs of modern immunology has been a clear resolution of this fascinating question. A wit proposed that this miracle of a process be referred to as the **Generation of Diversity,** or **GOD** for short, and this abbreviation stuck. I shall later give the briefest outline of how the GOD is achieved.

4. The attribute of self-nonself discrimination by innate defense and by the immune system

We will now consider how the innate system ensures that their mechanisms of attack are effective against foreign invaders whilst sparing host cells. All defense mechanisms must have this attribute of *self-nonself discrimination.* It is clearly important when phagocytes attack cells that they preferentially recognize and attack foreign cells rather than self cells. It is similarly important in acute inflammation that the protective molecules, brought to the site of injury, are more toxic for any invading, foreign microorganism than for cells of the host. This preferential targeting by phagocytes and toxic molecules of foreign rather than of self cells is the expression by innate defense mechanisms of the attribute of self-nonself discrimination.

Phagocytes have receptors complementary to structures present on invading microorganisms, but not on self cells. For example, there are components of the membranes of many and diverse bacteria, called *lipopolysaccharides*, that are not present on vertebrate cells. Phagocytic cells have surface receptors able to recognize and bind these lipopolysaccharides. In fact, lipopolysaccharides are just one of many *pathogen-associated molecular patterns*, or **PAMPs,**[15] uniquely present on foreign invaders and recognized by receptors on diverse self cells involved in defense, such as phagocytes. These receptors are called *pattern recognition receptors (PRR).*[15]

Diverse PRRs recognise PAMPs associated with diverse common structures characteristic of bacteria, of viruses, of protozoa, and of multicellular parasites, that are not present in or on vertebrate cells. For example, many bacteria have flagella protruding from their surfaces, which can rotate and are used as paddles, allowing the bacteria to swim. These flagella are composed of diverse proteins of which flagellin is the predominant component. The flagellin of diverse bacteria share certain structural features. There are receptors on vertebrate cells that recognize these common structural features of flagellin. These receptors, binding the PAMP that is flagellin, can initiate an attack on the bacterium. The presence of perhaps a hundred PRRs, recognizing a hundred different PAMPS commonly present on diverse invaders, might well provide a defense system that allows the vast majority of invaders to be attacked. Different PAMPS would thus represent different "flags" characteristic of different biological groups of invaders.

The ability of innate mechanisms of defense to target foreign invaders, but not self-cells, is due to the existence of PRRs recognizing PAMPs. These PRRs, almost always proteins, are coded for by the genes of the host. Thus, the self-nonself discrimination of innate defense mechanisms is said to be ***germline encoded***, i.e. it is a consequence of the specificity of the receptors that are coded for by genes handed down from parents to offspring in the usual way. We have seen that the immune system also has the attribute of self-nonself discrimination. Nevertheless, this attribute of the immune system has a radically different basis from the corresponding attribute of innate defense.

The universality of my immune system means that it can recognize the antigens of my parents, of my siblings, or of other people, distinct from my own, as foreign. Indeed, this is why there is a barrier to the successful transplantation of foreign organs, for example of kidneys between brothers. However, except in rare instances, we do not make damaging immune responses against the molecules and organs of our own body. The ability of the immune system to recognize and launch an immune response against foreign but not self antigens is referred to as ***immunological self-nonself discrimination.*** We have seen that Paul Ehrlich was the first to explicitly recognize this central attribute of the immune system in the early 1900s.[1] As he said, "The formation of (self) tissue (specific antibody) would… constitute a danger threatening the organism much more frequently and seriously than all exogenous injuries."

Efforts to understand the basis of self-nonself discrimination were central to the formulation of the Clonal Selection Theory of antibody formation, the basic theory in terms of which the functioning of the immune system is now analyzed. We shall delay a discussion of detailed ideas on how self-tolerance is achieved to the next chapter where the genesis of this theory is described. We just note here that the attribute of universality of the immune system explains why I can respond to my sibling's antigens as foreign. We have seen that this ability leads to the supposition that the immune system has the intrinsic ability to respond to what is self. Lack of response to self must therefore rely on a learning mechanism of what is self, ablating, or holding in check, this intrinsic ability to respond. Thus, the attribute of universality seems to require the attribute of a learned mechanism of self-nonself discrimination, leading to the suggestion that these attributes must have evolved hand-in hand. We shall also see, as we trace the considerations leading to the Clonal Selection Theory, that the attributes of specificity and universality led to an appreciation of why it is evolutionarily advantageous for the immune system to possess memory, an example of the attribute of adaptability.

5. The attribute of immune class regulation:
the choice of how to attack a foreign invader

How antibodies trigger the mechanisms of attack

Antibodies are proteins that have a highly variable part and a relatively non-variable or **constant part**. The highly variable part, which is responsible for the specificity of the antibody, binds the antigen. Antibodies that bind influenza virus are referred to as anti-influenza antibodies. As we have noted, antibodies are multivalent. When the antibodies bind to more than one antigen molecule, as can occur when antibodies bind to a virus particle, the antibody molecules can become distorted, or several antibody molecules can bind close together on binding to such a particle, so that they are effectively held close together in space and so appear to be aggregated. The resulting antibody-antigen complex can then be efficiently recognized by other molecular or cellular components of the body that are sticky for the distorted or aggregated, constant parts of the antibody molecules.

For example, phagocytes have receptors on their surfaces that can bind very weakly to the constant region of antibody molecules. This binding is so weak, though, that single antibodies rarely bind to these receptors. However, when a number of antibodies bind to a virus or bacterium, the resulting antibody-viral or antibody-bacterial complex becomes sticky for the surface of the phagocyte, as several receptors on the surface of the phagocyte can simultaneously bind to the complex. This binding initiates the envelopment by the phagocyte of the antibody-coated virus or bacterium.

Another example of an antibody-dependent mechanism of attack involves the interacting complex of molecules called **complement**. One component of complement can bind to antibody molecules that, in turn, are bound to the surface of a bacterium, and can initiate the piggy-back reaction that results in a hole being punched into the bacterial membrane, see Figure 2, and the consequent lysis of the bacterium.

Yet another type or class of antibody binds to mast cells and can mediate an antigen-dependent, acute inflammatory response. We have already seen how a scratch can activate pressure-nerve cells to release a substance that, in turn, activates mast cells, strategically placed next to these nerve cells, to release molecules such as histamine that locally trigger acute inflammation. The immune system has found a way of harnessing the process of acute inflammation in a highly specific manner. There are antibodies in individual people and animals that have been exposed to an antigen and subsequently become allergic to it. These antibodies bind to mast cells distributed throughout the body. When the antigen gains access to an internal site of the individual, the antigen interacts with the antibody bound to the mast cell's surface. The antigen thus aggregates these mast cell-bound antibodies, forming clusters. This clustering then signals the mast cells to locally release small molecules, such as histamine, to mediate acute inflammation at this site. Cells and fluid consequently rapidly accumulate at this site. This is the basis of the undesirable state known as **allergy**, which can lead to **asthma**, triggered by lung exposure to **allergens.** The allergic response is of course not always detrimental, and is important in providing protection against some extracellular parasites, such as helminths or worms.

In all these three cases, antibody-mediated phagocytosis, antibody-dependent, complement-mediated bacterial lysis, and antibody-mediated triggering of acute inflammation, the antibody molecules act as *connectors*. The antibodies bind through their highly variable part to the antigen/invader, whilst their relatively constant part interacts with elements of innate defense, such as complement, phagocytes, or mast cells, to bring the weapons of attack to bear upon the invader.

Physiological implications of the structure of antibodies

The role of antibodies, as connectors between foreign invaders and molecules mediating an attack upon the invaders, is critical in defending the host. We need to know more about the structure of antibodies to gain further insight into their role as connectors. This knowledge will allow us to appreciate further remarkable and important properties of the immune system.

Immunization sometimes leads to a notable increase in the amount of protein in the fraction of serum proteins known as γ globulins. Moreover, adding antigen to immune serum, resulting in the precipitation of antibody/antigen complexes, leads to a noticeable drop in the amount of protein in this fraction. It was therefore decided to call these antibodies **immune γ globulins**, abbreviated as **IgG.**

A series of discoveries led to the elucidation of the structure of these immune globulins in the 1960s to 1980s. All IgG antibodies are made up from a basic unit, consisting of two identical light and two identical heavy chains. This unit therefore has two identical halves. Each chain, whether a light or heavy chain, is made up of **domains** of about 120 amino acids, see Figure 7. The light chain consists of two such domains, the heavy chain of four or five. The first domains of both the light and heavy chain are highly variable, and are therefore called the variable domain, or V_L and V_H. They are contiguous in space and form the **antigen binding site** of the antibody molecule.[16] Thus this basic unit is divalent.

The second domain of the light chain does not vary and is referred to as the constant or C_L domain. The **constant region domains** of the heavy chains are referred to as C_H1, C_H2, C_H3, etc. There are a few different classes of antibody, eight in people, and the antibodies belonging to these different classes are different at the molecular level by virtue of the constant region of their heavy chains. The constant region of one type of heavy chain is unique in fostering the covalent joining together of five of the basic units, illustrated in Figure 7. This results in the largest type of antibody molecule, with a **valency** of ten (2x5=10). This type of antibody moleculee is known as **immune macroglobulin**, or IgM. The other important classes of antibody in humans are IgA, IgE, and IgG. The IgG class contains the IgG_1, IgG_2, IgG_3, and IgG_4 subclasses of antibody. These different classes and subclasses do not reflect mere details of the immune system, but are related to some basic physiological attributes. Antibodies belonging to these different classes act as connectors between the antigen and different weapons of attack. Thus, the nature of the attack depends upon the class of antibody present.

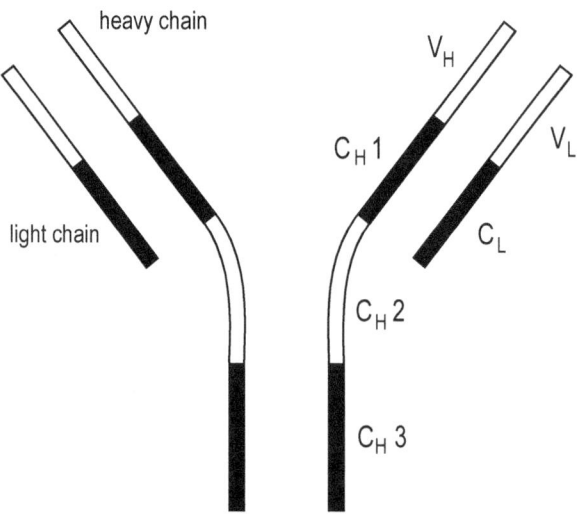

Fig. 07. The prototypical unit out of which all antibodies are constructed

The heavy chain of an antibody molecule is referred to by the Greek letter corresponding to the Latin letter that defines the class or subclass of the antibody. Thus *IgM* molecules have μ heavy chains, and *IgG_1* and *IgG_3* molecules have γ1 and γ3 heavy chains. *IgA* and *IgE* molecules have α and ε heavy chains.

Different classes of immunity are differentially regulated

An appreciation of the importance of there being several classes and subclasses of antibody came with the realization that antibodies belonging to different classes are usually optimally produced under distinct circumstances. Thus different effector functions are activated under these distinct circumstances. An infant allergic to cow milk and a young child that is allergic to peanuts have IgE antibody specific respectively for cow milk and peanut antigens. Antibodies belonging to the IgE class bind to mast cells. More often than not, infants grow out of their sensitivity to cow milk. This occurs as the class of antibody predominantly made against cow milk antigen changes from the IgE/IgG_1 classes to the IgA/IgG_4 classes, as we shall later discuss.

Similarly, some people produce substantial amounts of IgE and IgG$_1$ antibody to house mite dust, whereas others, living in the same environment, predominantly produce IgA and IgG$_4$ antibody. The former are *allergic*, whereas the latter individuals are said to be *non-allergic*, as IgA and IgG$_4$ antibodies do not bind to mast cells and so cannot trigger acute inflammation.[3] Such observations prompt two further questions: why do some individuals make one type of response and others make another? Perhaps even more important, what purpose does it serve the host to have these different means of attacking foreign invaders? We shall return many times to these interesting questions.

The observations just described illustrate two properties of the immune system that are highly significant. First, there are different *classes of immunity* that have different characteristics. Second, an individual's immune system does not generate all these different classes of immunity when it encounters antigen, but it has a *decision-making process* to decide which class(es) of immunity to produce. Moreover, the type of immunity can change during the course of an immune response, particularly noticeable when the response is prolonged, such as in AIDS. The mechanisms constituting these decision-making and implementing processes, which determine which class of immunity is generated, are collectively referred to as *immune class regulation.*

The discovery of cell-mediated immunity

The first realization that there are forms of immunity, other than those due to antibody, arose from Robert Koch's discovery that a bacterium, *Mycobacterium tuberculosis*, causes tuberculosis. In the late 1800s, Koch attempted to immunize patients with antigens derived from these bacteria, hoping in this manner to boost their immune response to the pathogen and thereby to resolve the disease. This treatment sometimes seemed to improve the patient's health and appeared, in other cases, to cause the patient to get rapidly sicker. Koch had to abandon this approach. Nevertheless, while attempting to develop the treatment, he made an important discovery.

A simple test of whether an individual is allergic to an antigen is to inject the antigen into the skin at a particular site, and see whether acute inflammation is produced there within minutes.[12] Such a reaction indicates what is called a state of *immediate hypersensitivity* and is due, as we have seen, to IgE antibodies, specific for the antigen and bound to mast cells. The impingement of antigen triggers the mast cells to release molecules, such as histamine, at the injection site, resulting in an inflammatory response at this site.

When Koch injected an antigen extract from *Mycobacterium tuberculosis* into the skin of tuberculosis patients, he did not see a reaction indicating a state of immediate hypersensitivity; rather, he saw a definite swelling around the site of injection, but this was only first apparent about twelve hours after the injection of antigen and was most apparent at 24-48 hours. This reaction came to be called ***delayed type hypersensitivity*** (**DTH**) to distinguish it from immediate hypersensitivity. This skin test, employing a ***purified protein derivative (PPD)*** of *Mycobacterium tuberculosis*, is known as the ***PPD*** or ***tuberculin skin test (TST)***.[13] A positive TST indicates an individual is infected by mycobacteria and is, if detected in conjunction with other clinical symptoms, used as evidence of infection by *Mycobacterium tuberculosis*. The skin test is a specific reaction in the same way as the precipitin reaction.

A state of DTH cannot be transferred from an immune to a naïve animal by the transfer of immune serum to the naïve recipient. This contrasts with the ability to transfer immune resistance to a bacterial toxin by administering immune serum to a naïve recipient, and the ability to transfer a state of immediate hypersensitivity, mediated by IgE antibodies. This inability led to the correct surmise that the DTH reaction is not mediated by antibodies. Certain developments had to occur before it could be clearly shown that the transfer of this type of immunity could only be achieved by transferring cells from an immune to a naive recipient.

It is difficult, in general, to stably transfer cells from one person or animal to another, as the recipient's immune system will usually recognize cells from another person or animal as foreign, and will therefore immunologically attack them. However, biologists in the first half of the twentieth century generated, by brother/sister mating over many generations, inbred strains of mice, guinea pigs and rabbits, in a manner similar to the breeding that led to inbred strains of dogs. All the individuals belonging to one strain, and of the same sex, were for all practical purposes genetically identical. The animals belonging to one strain are said to be ***syngeneic*** (Greek, same genes).

It is possible to transfer cells stably from immune animals to naive recipients, so long as they belong to the same inbred strain. The development of inbred stains allowed Landsteiner and Chase to demonstrate in the early 1940s that DTH is mediated by cells. Such a state of DTH can be transferred from an immune to a naïve syngeneic guinea pig by the transfer of spleen cells, but not of antibody. It was concluded that DTH is an expression of ***cell-mediated immunity***.[17]

Another form of cell-mediated immunity was discovered later. When individuals are infected by intracellular pathogens, viruses in particular, antigen-specific lymphocytes are usually generated that can preferentially recognize pathogen-infected but not uninfected host cells. These cytotoxic cells cause the lysis of the infected cell following this recognition. Such cells are called ***cytotoxic T lymphocytes***, usually referred to by the abbreviation ***CTL***.[18]

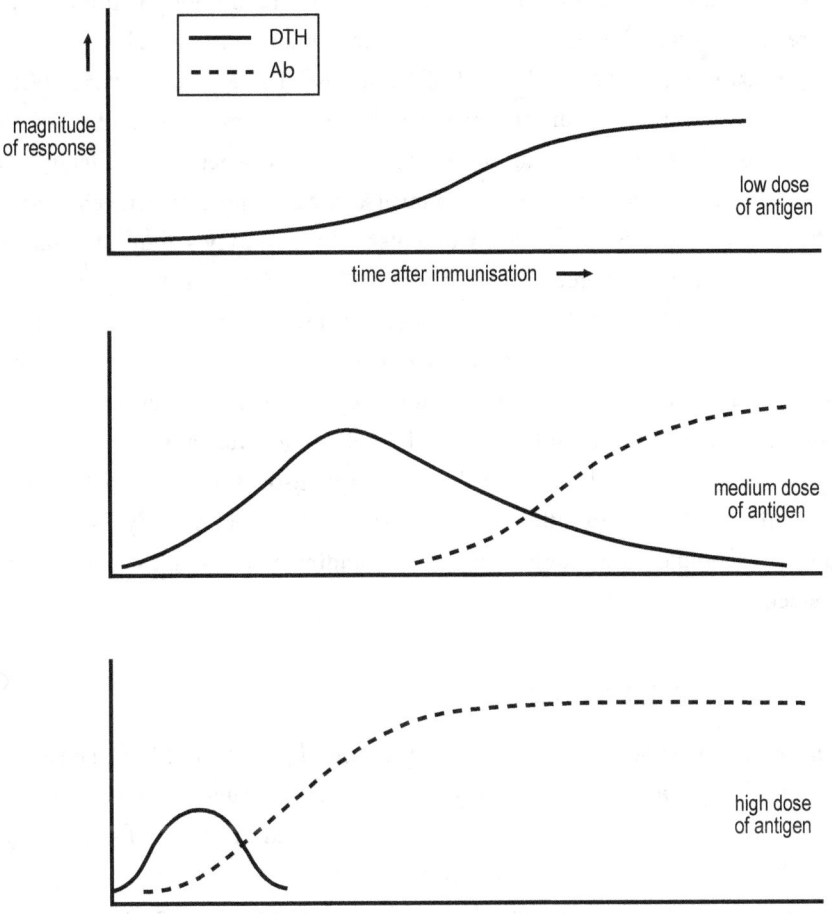

Fig. 08. The dependence of the kinetics of DTH and antibody
responses on antigen dose as described by Salvin[19]

The tendency for exclusivity in the generation of distinct classes of immunity

It soon became clear that different circumstances result in the optimal generation of cell-mediated and **humoral responses**, i.e. the production of antibody. For example, as we shall later discuss in more depth, some antigens can only generate cell-mediated responses. Other antigens can induce cell-mediated or humoral immunity, or even mixed responses, depending upon the circumstances under which the antigen encounters the immune system. However, even in cases where both these two major types of immunity ensue following immunization, there is a tendency for exclusiveness in their induction. Thus, Salvin showed in the 1950s that a DTH response is usually produced before antibody could be detected and, as antibody is produced, the expression of DTH usually declines. He demonstrated that, as the dose of antigen is increased, the tempo of the response also increases such that, if the dose of antigen was sufficiently increased, the expression of DTH rapidly becomes transient or may even be undetectable before antibody is produced, see Figure 8. Conversely, he found that it was possible to immunize with low amounts of antigen that resulted in an exclusive DTH response. These observations reflect the tendency for exclusivity in the generation of different classes of immunity.[19] We shall later discuss ideas on how the immune system "decides" whether to generate a cell-mediated or an antibody response. The inverse relationship between cell-mediated and antibody responses was first documented in a medically significant context in leprosy[20] and later in tuberculosis patients.[21]

Questions concerning the nature of these decision making processes have largely been formulated within the framework that there are two arms of the immune system, and that the immune system tends to optimally activate only one arm at a time. This framework is useful, but it is advisable to be aware at the outset of some limitations. There are, at least for the humoral arm, distinct branches that reflect the differential production of different classes or subclasses of antibody. We shall recognize this complexity as our growing understanding reveals that the two-arm view is inadequate.

Synopsis of Chapter 1

Immune systems exist only in vertebrates. They provide defense against foreign invaders by adopting older mechanisms of innate defense, also present in non-vertebrates. These older mechanisms include acute inflammation, phagocytosis, and the action of such molecules as complement and interferon. An appreciation of how the immune system adopts these evolutionary older mechanisms came through the study of immunity against pathogens.

Jenner immunized people in the late 1700s with material harvested from cattle suffering from cowpox to provide protection against smallpox. This success is a landmark of western medicine. The basis of this success became clear with the studies of Robert Koch and Louis Pasteur roughly fifty years later. They established that many diseases were due to infectious agents, i.e. to virulent pathogens. Pasteur developed different ways of cultivating or passaging virulent organisms so that they became attenuated. Infection with attenuated pathogens did not cause severe disease and also led to resistance to a challenge of virulent organisms. This resistance reflects the attribute of **immunological memory,** *a characteristic that distinguishes immune from innate mechanisms. Pasteur referred to this process of immunization against disease as vaccination.*

Resistance could be achieved either by active immunization or passively, by giving a naïve animal serum derived from an actively immunized animal. This latter finding was seminal and led to a characterization of the protective antibodies present in immune serum.

Immune serum, incubated with pathogenic bacteria at 37°C, leads to the disappearance of the bacteria within an hour. In this case, anti-bacterial antibodies bind to both the bacteria and to a component of complement. This interaction initiates the complement-mediated process of the formation of holes in the membranes of the bacteria (Figure 2), and so to bacterial death. In this way, the immune system adopts the innate mechanism of defense known as complement.

The pathogenesis of some bacteria is mediated by their production of toxins. Immunization with toxoid, a slightly altered and non-toxic form of the toxin, protects against a subsequent challenge of the toxin. When toxin-immune serum is added to a transparent solution of the toxin, the mixture turns cloudy and a white precipitate forms. This process is known as the precipitin reaction. Marrack proposed the Lattice Hypothesis to explain the phenomenon. Both antibody and antigen were envisaged to be multivalent. The presence of roughly equal numbers of antibody and antigen molecules would result in the formation of large nets or lattices of linked antigen and antibodty molecules, see Figure 4. Marrack proposed that large lattices are insoluble and so precipitate. Compelling evidence supports his hypothesis.

The next breakthrough was the recognition of how extremely specific antibodies are in their binding to antigen. In the early 1900s, investigators explored whether antibodies could be raised against small, organic molecules of known structure. Antibodies to a small molecular hapten can be raised by chemically conjugating it to an immunogenic carrier, and immunizing with the conjugate. Lansteiner showed that such antibodies could bind to the hapten but not to molecules very closely related in structure to the hapten. These studies demonstrated how extremely specific antibodies can be.

Landsteiner's studies not only demonstrate the exquisite specificity of antibodies, but also that antibody can be produced against virtually any foreign molecule greater than a certain size. The ability to respond to all foreign molecules is referred to as the attribute of **universality**. Around this time, Paul Ehrlich found that antibody could be raised in one goat against the red blood cells of another goat, and that this antibody did not react with the red blood cells of the immunized goat. This led Ehrlich to speculate on the existence of an attribute later referred to as **self-nonself discrimination**. He envisaged that the immune system has a means to ensure that antibodies are not produced to "self antigens".

The last attribute of immune class regulation became apparent later. Immunity can be mediated by cells or by antibody, respectively referred to as cell-mediated immunity and humoral or antibody-mediated immunity. In addition, there are different classes of antibody. The optimal generation of cell-mediated immunity, and optimal production of different classes of antibody, occurs under different conditions. What is the biological significance of this differential regulation and how might such regulation be achieved? These questions are directed at understanding the attribute of **immune class regulation**.

The two attributes of specificity and universality must mean there is an enormous repertoire of different antibody molecules that an animal or person can produce. How this enormous number of different antibodies could be generated, and how self-nonself discrimination could be achieved, became the two dominant questions from the 1920s to the 1960s. Attempts to answer them drove much of the most interesting speculations in the *field and gave rise, after several twists and turns and through a collective effort of several individuals, to the Clonal Selection Theory of antibody formation.*

Chapter 2

THE CLONAL SELECTION THEORY
OF ANTIBODY FORMATION

Overview

The Clonal Selection Theory was developed in the 1950s and 1960s and provides the central framework of the field. Our immunological speculations and discussions will take place in the context of this theory. I, and the large majority of immunologists, are convinced of its essential validity. The purpose of this chapter is to convey to the reader not only the circumstances that led to this theory, but why it is believed and why it constitutes the cornerstone of the subject. I am indebted in giving this description of the birth of the Clonal Selection Theory to the introductions of two PhD theses, in which Calliopi Havele[22] and Nathan Peters,[23] then graduate students, provided thoroughly researched and scholarly accounts of the events surrounding these critical developments in the history of our discipline.

Ideas leading up to the formulation of the Clonal Selection Theory

The first known and documented theory of how antibodies are produced was made by Ehrlich, and described in an address in 1901 to The Royal Society of London.[24] He proposed that there are cells, each with many different kinds of receptors on their surface. He envisaged this class of receptor might have had a primeval purpose of transporting small molecules from the outside to the inside of the cell, these small molecules being required for the cell's sustenance and growth. Ehrlich proposed that when antigens bind to similar receptors, the cell bearing the receptor is stimulated to produce and secrete molecules of the same specificity as the receptor with which the antigen interacted. He proposed that these secreted molecules accumulate as antibodies in the blood. Ehrlich's was the first **Selective Theory** of antibody forma-tion, in the sense that the antigen, by binding to particular receptors, selects the kind of molecule to be produced.

As we have seen, the subsequent work of Landsteiner in the early decades of the 1900s, on raising antibodies to structurally well-characterized haptens, led to an appreciation of the astronomical number of different antibodies an animal can make. This led to a serious questioning of Ehrlich's selective view, which meant that the full diversity of antibody molecules had to exist in the animal before it was immunized with a particular antigen. Some thought this unbelievable, incredible in the literal sense, and were forced to consider the alternative: antibodies do not pre-exist the impingement of antigen upon the immune system, but rather the antigen is employed in some manner to make, or instruct, the formation of an antibody complementary to the antigen. These theories of antibody formation were collec-tively called **Instructive Theories**. The well-known immunologist Niels Jerne was responsible for an everyday analogy that nicely illustrates the difference between Instructive and Selective Theories. A tailor that makes a jacket to fit an individual takes measurements, and cuts and sows the cloth accordingly. Here the individual is held to be analogous to the antigen and the jacket to the antibody. The tailor is the mediator of the instructive process of obtaining a jacket/antibody complemen-tary to the individual/antigen. However, most of us go to a big department store, which has a wide selection of pre-made jackets, and we select one that fits us well enough. This example illustrates the most important difference between Instructive and Selective Theories. According to Instructive Theories, the antigen is required to generate a complementary antibody; according to Selective Theories, the antibody must exist before the antigen can select it.

A weakness of most Instructive Theories was the lack of any plausible mechanism by which an antigen could instruct the formation of a complementary antibody. This vagueness was inevitable, as there was no knowledge at the time as to how macro-molecules, such as antibodies, are made. However, such a weakness should not be considered fatal, as the most interesting questions are usually posed in the context of the relatively unknown. Linus Pauling published his influential Instructionist Theory in 1940, made in the context that antibodies were then recognized to be protein mol-ecules, consisting of one or more polypeptide chains. Pauling suggested these chains were intrinsically flexible and wrap themselves around antigen molecules, thereby defining the antibody's eventual shape. He envisaged that the antibody would now preferentially bind to the "template-antigen".[25]

Burnet and Fenner listed in 1949 some phenomena that they considered dif-ficult to understand in the context of Instructionist Theories.[26] This publication is truly fascinating as an example of pivotal ideas, some correct and others wrong, that attempt to address major questions. First, a hallmark of the antibody response is the existence of memory; secondary responses are usually greater and more rapid than primary responses. What could be the basis of memory in the context of Instructionist Theories? It was difficult to conceive of a basis. Second, there is a lag period of a few days after antigen impinges upon the immune system before substantial antibody is apparent, after which its appearance increases in an escalating fashion. How could the lag period and this increasing tempo of antibody production be understood in the context of Instructionist Theories? Moreover, it would appear that there are many more self than foreign molecules in the body at any particular time, so should not these self molecules be instructing the formation of the majority of antibodies to be complementary to themselves? Should we not expect, according to Pauling's model, that the vast majority of the flexible polypeptide chains would fold up around self molecules rather than the relatively few foreign molecules present? How could the attribute of self-nonself discrimination be understood in the context of these theories?

Burnet and Fenner made a first and partial proposal as to how self-nonself discrimination might be achieved. They pointed out that self but not foreign antigens are present during embryonic development. They proposed that all self antigens were somehow biochemically marked as self at this time, and that the existence of this marker blocked the subsequent formation of antibody against the marked antigen. The idea, that the presence of self antigens is required early in the development of an animal to prevent the immune system from responding to these antigens later on, was supported by the subsequent experiments of Billingham, Brent, and Medawar. They published their findings in 1953.[27]

Mice, exposed neonatally to cells from another strain, could not as adults reject skin from this donor strain. They could, however, reject skin derived from other strains. The authors interpreted this to mean that the presence of foreign antigens during development resulted in the immune system regarding the foreign antigens as self, and were therefore unable to subsequently respond against them. The mice were *tolerant* of the donor antigens. This unresponsiveness reflected a *state of tolerance*, or more shortly, of *tolerance.* These observations validated Burnet and Fenner's idea that the presence of a self antigen during development, i.e. during the animal's early history, is critical to the immune system regarding the antigen as self. We shall refer to this idea as the *Historical Postulate.* Medawar and Burnet subsequently received a Nobel Prize for their work on tolerance. However, the idea that self antigens were modified during embryonic life to express a self-marker, that inhibited the process of antibody formation, blocking the generation of immune responses to self antigens, did not stand the test of time.

The impact of molecular biology on the
birth of the Clonal Selection Theory

By the time that the debate between selective and instructionist theories reached its peak intensity in the 1950s, it had become firmly established that antibody molecules are proteins. Molecular biology arose as a recognized discipline around the same time. Evidence had shown that DNA is the genetic material, and that it somehow also determines the structure of proteins. Central to molecular biology was the key question of how DNA specifies the nature of proteins. The answer was likely to be pertinent to the origin of antibody molecules, unless antibodies are a special case. The enormous number of antibodies an individual can produce was beginning to be appreciated at this time. Crick, in discussing the general question of how DNA might code for polypeptide chains, raised the possibility that antibody chains might be an exception to the general considerations he was putting forward. This probably reflected a concern over how the available DNA could code for so many antibody molecules.

Jerne proposed a selective theory in 1955,[28], probably the first since Ehrlich's in 1901. During his PhD studies, Jerne had found that rabbits, not deliberately immunized, had in their serum low but detectable levels of antibodies against the antigen he used. This led him to suggest that the presence of the antigen is not needed to make the corresponding antibody, but rather that the antigen selects complementary, pre-existing antibodies. His theory posited three steps. (i) A diverse array of antibodies is made early in development, and these antibodies are present in the blood. (ii) Those antibodies, specific for self antigens present at this early time in development, combine with the self antigens and are removed. These two processes leave a diverse population of antibodies in the blood of a developing person or animal, with the potential to bind to foreign but not to self antigens. (iii) Jerne's proposed third step was that, when a foreign antigen impinges upon the body of a mature individual, the antigen binds complementary antibody. The resulting antigen/antibody complex is taken up by a phagocytic cell, where the antibody directs the cell to produce more antibody molecules identical to itself.

Jerne's theory was indeed a selective theory, in the sense that antigen selects pre-existing antibody for replication. However, the process by which antibody could replicate itself, or a protein provide instructions for the synthesis of a protein identical to itself, as he envisaged in the third step, was obscure. His proposal was difficult to reconcile with the recent insights that a particular piece of DNA specifies the nature of a particular polypeptide chain. This aspect of his hypothesis had the aroma of instructionist theories. I remember reading somewhere that Jerne tried to get Jim Watson's reaction to his ideas. Jim, famous with Francis Crick for elucidating the structure of DNA, was true to form, and not polite. He said, if I recall correctly, that the theory stank. Nevertheless, Jerne's theory was historically most important.

Burnet no doubt contributed considerably to the ideas that came together to form the coherent framework now known as the Clonal Selection Theory.[29] However, understanding what he wrote is often not easy, and his writing is often obscure when one compares his expositions to those of his contemporaries. It is my feeling that Burnet was perhaps not as analytical as others, but imaginatively faced big questions without intimidation. He championed both incorrect as well as novel and important ideas, and demanded less than others that published ideas are clear and consistent within a plausible framework. In 1949, for example, in discussing possibilities within an instructionist framework, he thought that somehow the instructions might affect a cell so that it, and its descendants, could produce antibody complementary to the antigen. This instructionist theory was again somewhat obscure from a mechanistic point of view, and clearly different from Pauling's flexible chain theory. This proposal, nevertheless, was a first attempt to understand the basis of immunological memory, and it resurfaced as a central feature of Clonal Selection Theory: different cells can have different, inherited potentials, so that these potentials are handed down from one cell to their offspring.

David Talmage made a considerable contribution at this juncture. One gets the impression through his writings that he was much more careful, sparing, and considered than Burnet, who was sometimes widely speculative. Talmage appreciated the deeper issues being faced by those establishing the foundations of molecular biology. Proteins do not provide the instructions for their own replication, as Jerne's hypothesis demanded. Talmage proposed in 1957 that "one of the multiplying units of the antibody response is the cell itself...only those cells are selected for multiplication, whose synthesized product has affinity for the antigen injected. This would [require] a different species of cell for each species of protein produced..." Talmage also included the idea that exposure to an antigen early in life would ablate the corresponding cells. The ability to produce antibody complementary to these antigens would be lost. He provided in this manner a potential explanation for Medawar's observations on tolerance and for how self-nonself discrimination might be achieved.[30]

A feature of Talmage's proposal is worth stressing. Given the great specificity of antibodies, and the universality of the immune response, the cells with antibody receptors specific for a particular antigen, before this antigen first impinges upon the immune system, must be very scarce indeed. As indicated in the above quotation, "those cells selected for multiplication", Talmage envisaged that the cells producing antibody arose through multiplication of "precursor cells". This idea was indirectly supported by two observations. There is a lag period of several days following immunization before antibody can be detected, and then the amount produced rises very fast. This lag phase was naturally explained by a period during which the stimulated but initially scarce precursor cells divided. In addition, it had been found that the ability of an animal to produce antibody against an antigen was highly sensitive to radiation of the animal. Cells exposed to radiation and forced into cell division usually die, due to damage of the few essential DNA molecules each cell contains. The radiation sensitivity of the immune response is readily understood if, during the course of an antibody response, the scarce precursor cells, when activated, multiply before they give rise to antibody-producing cells.

Talmage's proposal was clear, short, and succinct.[30] It seems to be the first published and coherent account of the essential elements of the Clonal Selection Theory. Nathan Peters, in the introduction of his PhD thesis, attempts to provide a judicious account of the influences of the different contributors upon one another:

> In order to appreciate the contributions of Ehrlich, Jerne, Talmage, and later Burnet, who developed the idea of the Clonal Selection Theory to its logical conclusion, it is necessary to determine the degree to which these authors influenced each other. Burnet had implied that cell proliferation may be involved in immune responses as early as 1949, when he stated 'that antibody production is a function not only of the cells stimulated, but of their descendants.' Yet as late as 1956 he still maintained that antigen directed, rather than selected for, the formation of antibody. In contrast, Jerne suggested in 1955, as had Ehrlich in 1901, that the antigen selected the receptor.[23]

The Clonal Selection Theory becomes the foundation of immunology

Molecular Biology and the Sequence Hypothesis

There is no doubt that the establishment of molecular biology in the 1950s to the 1970s had a profound impact on immunology. I shall summarize briefly the major ideas of molecular biology that are necessary to appreciate its impact on immunology.

DNA is the genetic material and so responsible for the inheritance of traits. Observations had shown that DNA carries out this hereditary function by, for the most part, specifying the nature of the proteins that the cells of an animal produce. The DNA molecule itself consists of two entwined chains. These have direction in the sense that the two ends of a given chain are dissimilar, so that one end can be usefully defined as the beginning and the other as the end. The two entwined chains of a DNA molecule run in opposite directions. They are themselves made from linking together four bases, abbreviated as A, T, C and G, in a regular manner. Thus the chemical nature of a DNA chain can be defined by the order of its bases, e.g. AACGCCTT. Watson and Crick's structure of DNA is based upon the rule that if one chain has A at a particular position, the entwined chain must have T at this position, and if one chain has C at a position, the entwined chain has G at this position. The sequence of one chain thus defines the sequence of the other. These rules of base pairing have a simple, structural basis: only A fits with T, and only C with G. This base-pairing rule provided the underlying idea for Watson and Crick's proposal for how DNA replication (duplication) occurs. The two entwined chains come apart, and each acts as a template to define its complement, and so one double-stranded parent molecule gives rise to two double-stranded daughters.[31] This mechanism can be appropriately seen as an instructive process.

Many of the structural components of a cell and its enzymes are proteins. Moreover, the other types of molecules are either obtained directly from food or are produced through the action of enzymes, most of these also being proteins. Thus, defining the proteins of a cell defines most of its essential components. A given protein consists of one or a few polypeptide chains. These polypeptide chains have a direction, in the sense that the two ends of the chain are different. Sanger first showed that a given polypeptide chain is made from linking together amino acids in a regular manner.[32] There are twenty different amino acids that are found in polypeptide chains. Sanger showed that each of the two polypeptide chains of insulin has a unique amino acid sequence. Let us for convenience refer to the twenty amino acids as "a" through "t". The chemical nature of a polypeptide chain is defined by specifying the order in which its amino acids are strung together, e.g. d.t.m.a.a.q.p

Classical genetics and biochemistry had led to the idea that a segment of DNA, called a gene, codes for a polypeptide chain. What was meant by the term "code" was obscure. The unraveling of the process that constitutes coding is one of the major triumphs of molecular biology.

It was known that the function of proteins, including enzymes, is critically dependent on their three-dimensional structure. How could the genetic material determine this structure? A central idea was that the genetic material defined the three dimensional structure of a protein indirectly. The **Sequence Hypothesis**, most clearly enunciated by Crick, stated that the genetic material only defines the order of the amino acids of a polypeptide chain, not its three dimensional structure. The polypeptide chain, as it is made, somehow folds up automatically to achieve the three-dimensional shape critical for its function. The sequence hypothesis was in time strongly supported by experiments, as we shall see.

The sequence hypothesis states that the particular sequence of bases of one of the two entwined strands of a stretch of DNA, the "sense strand", constituting a gene, determines the particular order of amino acids in the polypeptide chain that the gene encodes. George Gamow focused attention on how this relationship is achieved by making the first detailed proposal as to the nature of this genetic code. Molecular biologists subsequently elucidated how a sequence of DNA bases determines the sequence of amino acids of a polypeptide chain in considerable detail. This mechanism of protein synthesis, though fascinating and of fundamental importance, is not essential to our needs here.

What was the broad impact of these developments on immunology? First, strong evidence was found to support the Sequence Hypothesis for the proteinaceous enzyme ribonuclease. The Sequence Hypothesis stated that the polypeptides of proteins fold up into their critical three dimensional structure without a guiding mold, if they are present in a liquid that closely mimics the salty water that constitutes the fluid of the interior of cells and of the fluid in which red and white blood cells are suspended in our blood. It was known that, when this fluid is replaced by one with different properties, containing for example detergents, the polypeptides lose their original structure. This process is called **denaturation**. Anfinsen took ribonuclease, dissolved in salty water, and replaced the salty water with another fluid containing urea, leading to a fluid with different properties. He demonstrated that the enzyme lost its structure, i.e. was denatured. Anfinsen then returned the polypeptide chains of ribonuclease into a salty water environment, and the polypeptides again took up their critical three-dimensional shape, a process naturally called **renaturation**. Thus no mold was required for the polypeptide chains to regain their critical three-dimensional shape and their enzymatic activity.[33] These observations strongly support the Sequence Hypothesis. We shall shortly see their pertinence for immunology.

Antibodies and the Sequence Hypothesis

In trying to experimentally determine whether the Sequence Hypothesis applied to antibodies, it was deemed important to have a molecularly homogenous population of antibody molecules. When people examined antibodies, highly purified for their ability to bind antigen, they found that there were usually hundreds of chemically distinct molecular species. An alternative approach had to be found to obtain chemically homogeneous antibody molecules and avoid the complications that would arise from employing a heterogeneous population of molecules.

Some mice and humans develop cancers of antibody producing cells, called **multiple myeloma**. All the cancer cells from a given patient, or from a mouse with multiple myeloma, produce the same antibody, which accumulates in large amounts in the blood and so can be readily purified. People purified a number of these **myeloma proteins**. Two kinds of observation convinced most that the Sequence Hypothesis also applied to antibody molecules. Investigators found that different myeloma proteins bound different antigens and that the amino acid sequence of their constituent polypeptide chains varied among myeloma proteins. This was not anticipated in the context of Pauling's Instructionist Theory, according to which the same flexible chains could fold around different antigens to produce different antibodies binding to these different antigens. Moreover, myeloma proteins could be denatured in a similar fashion to that employed by Anfinsen in his studies with the enzyme ribonuclease, and subsequently renatured to retrieve their ability to bind their antigen. These processes took place in the absence of the antigen. This led to the inevitable conclusion that antibodies are like other proteins, in that the amino acid sequence of their constituent polypeptide chains determines their three dimensional shape, and thus their ability to bind antigen. The antibody chains did not need antigen to guide their folding.

The critical elements of The Clonal Selection Theory

In my opinion, the most complete, succinct, and elegant formulation of this theory was presented in 1959 by Joshua Lederberg in an article entitled "Genes and Antibodies"[34] in the magazine *Science*. The account given here is based on his statement of the theory. Lederberg's exposition was made in the context of antibody production, and we shall follow him in this respect. Nevertheless, we have already acknowledged the existence and importance of cell-mediated immunity. We shall later see that the Clonal Selection Theory also applies to the cells responsible for this form of immunity.

There are two central ideas at the core of this theory, and two additional ideas proposed and discussed by Lederberg and others. The first is that antibodies preexist the antigen. The second is that there is a diversity of precursor cells, each bearing a unique set of active genes encoding a unique antibody receptor; the interaction of antigen with this receptor can lead to the activation of the precursor cell, resulting in its multiplication and the production and secretion of antibody by its descendants. The specificity of the antibody produced was postulated to be identical to the specificity of the antibody receptor that the precursor cell bore.

These two postulates represent the core of Clonal Selection Theory. However, Lederberg and others went considerably further. Lederberg speculated, to good effect, as to how this diversity of precursor cells arose. Most thought at that time that all genes encoding polypeptides chains were handed down from parents to offspring. Most thought this must also be true of the genes coding for antibody light and heavy chains. This theory is, for obvious reasons, called the ***Germ Line Theory*** for the origin of antibody genes. It implies that these genes were generated in evolutionary time.

Lederberg and some others proposed something quite radical: they envisaged that antibody genes are generated during the lifetime of the individual, i.e. somatically, in contrast to being generated in evolutionary time. What are inherited are not the genes that code for light or heavy chains, but rather *kits to assemble such genes*. The assembly process would contain several steps with an arbitrary component, resulting in the possibility of many different genes being assembled in many different cells. Lederberg thus provided a vivid picture of how precursor cells might arise, each committed to making antibody of just one type. His insights have been shown to be valid, as we shall later see. This vision can be succinctly expressed by saying that the Generation of Diversity (of antibody genes), or GOD, occurs somatically, i.e. during the lifetime of an individual. This is the ***Somatic Theory for the GOD.***

This envisaged scenario led to the recognition of the pressing nature of yet another problem. How could self-nonself discrimination be achieved? It seemed inevitable that the proposed somatic process, underlying the GOD of immunoglobulin genes, would lead to the birth of precursor cells with antibody receptors able to bind self antigens. All contributors to the formulation of the Clonal Selection Theory recognized this problem, and all attempted to consider some principles by which self-nonself discrimination might be achieved. All except Lederberg accepted the Historical Postulate, as originally formulated by Burnet and Fenner, namely that tolerance to self was uniquely established in the developing fetus. However, in this case, how could autoimmunity arise in adults, as sometimes happens? Lederberg proposed a dynamic model in which new precursor cells arise throughout life. He thus envisaged that there was a continual need for eliminating new precursor cells with anti-self receptors. We shall refer to the Historical Postulate in the broader sense depicted in Figure 9 and as envisaged by Lederberg: tolerance to a self antigen requires its presence early in development and its continued presence thereafter.

Much contemporary discussion takes place in the context of Lederberg's hypothesis as to how self-nonself discrimination is achieved. Lederberg proposed that, when a cell first acquires an immunoglobulin receptor through the somatic GOD, the cell is programmed to die if its receptor interacts with antigen. This cell is thus in an ***inactivatable*** or ***paralysable state***.

If no such interaction with antigen occurs, the cell will, due to an internal clock, differentiate after a few days into a cell bearing the same receptor but now programmed to be ***activatable*** or ***inducible*** see Figure 9. Interaction of this inducible cell with antigen was envisaged to result in its multiplication and the differentiation of its descendants to produce antibody. Lederberg envisaged that all, or at least most, self antigens first appear early in development and are continuously present thereafter. Thus, precursor cells that are generated with an antibody receptor, able to bind a self antigen, will be inactivated when they first arise, and will not lead to precursor cells that can be activated. Those precursor cells born with a receptor not recognizing any self antigens will not be inactivated, but will give rise to a pool of cells whose receptors can recognize only foreign antigens. Some of these precursor cells may subsequently interact with a foreign antigen and be activated when such a foreign antigen impinges upon the immune system. In summary, Lederberg's proposal explained how activatable precursor cells could arise that had receptors able to bind foreign but not self antigens. We shall later come to appreciate how insightful this proposal is.

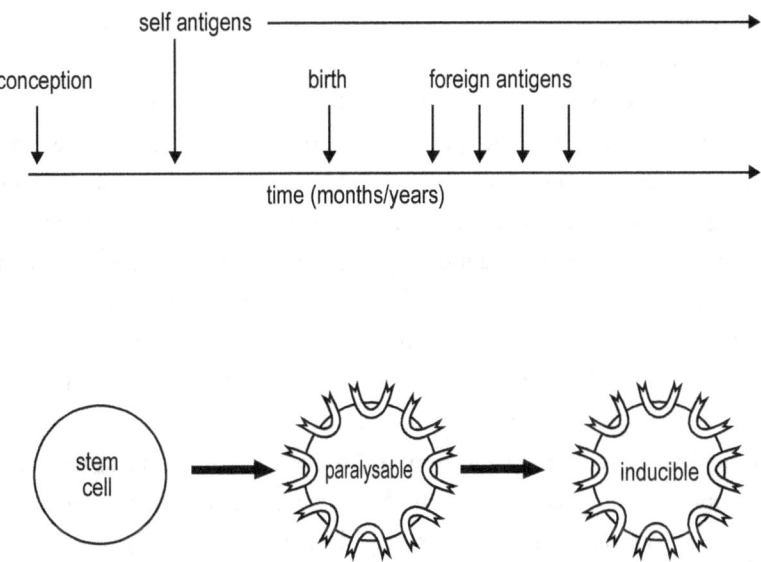

Fig. 09. Illustration of the Historical Postulate and of Lederberg's proposal for how self-nonself discrimination is achieved.

Functional tests of the Clonal Selection Theory

The clearest prediction of the Clonal Selection Theory is that there are different precursor cells, each bearing different antibody receptors able to interact with different, non-crossreacting antigens. When antigens interact with their respective precursor cells, the antigens can induce an antibody response. Thus, the removal from an animal of precursor cells able to bind to one antigen should affect the ability of the animal to respond to that antigen but not to a non-cross-reacting antigen. The truth of this proposition has been demonstrated in many systems employing the same underlying logic. In order to achieve this demonstration, it is necessary to have a source of precursor cells that can be activated by antigen to produce antibody. What could be a source of such cells?

We have seen that an animal's ability to generate an immune response is sensitive to whole body radiation. We have also seen that some immunologists seized upon this observation as consistent with their notion that the generation of an immune response required the antigen to activate scarce precursor cells to multiply and their descendants to differentiate to produce antibody. Irradiation would damage the DNA of precursor cells, whose aborted division and death upon being activated by the antigen would prevent them from giving rise to antibody-producing cells. If this surmise is true, it should be possible to reconstitute the irradiated animal's ability to make antibody responses by providing it with a source of unirradiated precursor cells obtained from a genetically identical animal.

Irradiated mice given syngeneic, unirradiated spleen cells and challenged with an immunogenic antigen vigorously produce antibody to the antigen. This observation led to the tentative conclusion that spleen cells contain precursor cells. This surmise was strengthened by a further observation. It was possible to show that about 30% of the white cells in the spleen of a mouse had on their surface mouse antibody molecules. This was expected if precursor cells bore antibody receptors. Rabbit antibody, able to bind to mouse antibodies, can be raised by immunizing rabbits with mouse antibodies. The rabbit anti-mouse antibody can be used to remove cells present in a population of mouse spleen cells that express on their surface mouse antibody receptors. Depletion of these cells from a spleen cell population abrogates the ability of this population to restore the antibody response when given to irradiated mice. These observations are consistent with the spleen containing precursor cells bearing antibody receptors on their surface.

It is possible to incubate spleen cells with a foreign antigen and detect rare cells that bind the antigen. Moreover, it is also possible to remove those cells binding the antigen, leaving the other cells behind. For example, an antigen with which such studies have been carried out is *sheep red blood cells (SRBC)*, as these foreign cells provoke a vigorous antibody response when injected into mice. Spleen cells from normal mice, when given to lethally irradiated syngeneic mice, allow the irradiated, reconstituted mice to make an antibody response when they are challenged with SRBC, but mice reconstituted with spleen cells, depleted of SRBC-binding cells, are unable to do so. Irradiated mice reconstituted with spleen cells depleted of SRBC-binding cells can make antibody responses to other antigens, non-crossreactive with SRBC, such as chicken RBC, when challenged with the appropriate antigen. These observations demonstrate that there are different precursor cells specific for SRBC and for chicken RBC, as envisaged in the Clonal Selection Theory.

Red blood cells (RBC) from different vertebrate species were often used as antigens in classical studies, as they generally induce brisk and strong immune responses without the aid of *adjuvants.* In contrast, purified foreign proteins obtained from different vertebrate species often do not generate strong immune responses unless given with adjuvant. Adjuvants are usually a mixture of substances that, when given together with antigen, increase the size and rapidity of the immune response to the antigen. The use of adjuvants brings unknowns into how the observations are to be interpreted, a consideration favoring the use of xenogeneic RBC as antigens.

A note on terminology

Immunologists at the time Clonal Selection Theory was formulated normally referred to the cell that initially interacts with the antigen, and that gives rise to anti-body-producing cells, as the ***precursor cell*** or the ***antigen-sensitive cell.*** These terms are rarely used today unless talking in an historical context. Also, immunologists at this time primarily discussed their ideas in the context of antibody production, even though the existence of cell-mediated immunity was recognized. The reason for this is that more was then known about antibody than cell-mediated responses. However, we now know that Clonal Selection Theory applies to the cells involved in both types of response. In addition, we know that precursor cells or antigen-sensitive cells belong to the population of cells, characterized originally on morpho-logical grounds, as lymphocytes. In modern terminology, the old precursor cell is an antigen-specific lymphocyte. However, the correspondence between precursor cell and antigen-specific lymphocyte is inaccurate in at least two respects. First, there are different types of precursor cell, such as those that give rise to the cells that produce antibody and those that give rise to cells that mediate cell-mediated immunity. In contrast, the initial precursor cell was envisaged as a precursor cell for antibody for-mation. Second, the term "precursor cell" was originally employed to refer to a cell that had not interacted with antigen. With time, immunologists felt a need to make further distinctions, and talked of "virgin" or "naïve" precursor cells to distinguish them from memory precursor cells, responsible for secondary immune responses. When we now refer to antigen-specific lymphocytes, we mean a population of cells that contains different types of precursor cells, as well as the cells they give rise to on activation. All lymphocytes share with the old "precursor" or "antigen-sensitive" cell the production of just one kind of antigen-specific receptor. I shall henceforth often use the term lymphocyte in the sense indicated here, so that my terminology better reflects current usage.

Synopsis of Chapter 2

The Clonal Selection Theory was formulated to explain how an individual could both produce so many different antibody molecules, and could respond to foreign but not to self antigens.

The two competing ideas as to the origin of antibodies were referred to as Selective and Instructive Theories. Selective Theories postulate that those antibodies produced upon immunization must pre-exist the impingement of antigen upon the immune system. The antigen selects the pre-existing antibodies that are produced. Instructive Theories postulate that antigen is required for the de novo production of the corresponding antibody molecules. Two features of the antibody response were used to argue against instructive theories: the existence of a lag period of several days before antibody could be detected, and thereafter its appearance in an escalating manner; second, the existence of secondary antibody responses. It was unclear how either of these features could be accounted for by instructionist theories, as there was no obvious scope within an instructive framework for how "memory" could be generated. In addition, the problem of how self-nonself discrimination could be achieved was addressed by those espousing selectionist views, but not by those espousing instructionist ideas. Indeed, it is difficult to understand within an instructionist framework why the immune system is not pre-occupied with producing antibody to self antigens, which are collectively much more prevalent than foreign antigens.

It took time before the attractions of the Clonal Selection Theory were appreciated. Central to this appreciation was Burnet's early suggestion that the lag period seen in antibody responses would be understandable if substantial cell multiplication was required before measurable amounts of antibody could be produced, and if the characteristics of parental precursor cells to produce a particular antibody were handed down to their progeny. The faster tempo of secondary responses could then be envisaged as due to an increased frequency of the responding cells. Burnet and Fenner proposed that self antigens, present early in development, resulted in an ablation of the animal's ability to respond to self antigens. Medawar and colleagues experimentally confirmed this idea in 1953. They demonstrated that neonatal type A mice, given cells from type B mice, were able to accept B type skin grafts as adults.

During his graduate studies, Jerne found antigen-specific antibodies in non-immunized rabbits, a finding he took to be consistent with selective theories. Jerne proposed in 1955 the first Selective Theory of modern times. He postulated that early in their development, animals produce a diversity of antibody molecules. Those that react with self antigens are removed, leaving a population of antibodies able to bind to only foreign antigens. Jerne proposed that, when a foreign antigen impinges upon the immune system, antigen/antibody complexes form in the blood. These complexes are taken up by a macrophage, where the antibody is replicated. This proposal, that proteins could provide cellular machinery with the instructions for their own reproduction, ran counter to the ideas then being articulated in the new discipline of molecular biology.

In 1957, Talmage added to the interesting ideas of others the critical thought that the unit selected by antigen is a cell committed to the production of one antibody, by virtue of the particular antibody genes it expressed. The selected cell multiplied, and its descendants produced antibody of the same specificity as the antibody receptor of the selected precursor cell. Talmage also incorporated the idea that tolerance to self was established early in development as a result of the interaction of precursor cells with self antigens, leading to their ablation. Talmage's proposal was outlined in a couple of paragraphs in a review, which was a modest way of outlining what probably is the first coherent description of the Clonal Selection Theory.

Burnet subsequently wrote a book upon the subject. Lederberg gave the most careful, explicit, bold, beautiful, and axiomatic account of Clonal Selection Theory in an article in Science in 1959. He couched his exposition in terms of insights gained through the establishment of molecular biology and the analysis of the nature of bacterial mutations. He added one critical idea. Lederberg suggested that self-nonself discrimination is a continuous process that occurs throughout an organism's life, and proposed a simple mechanism, outlined below, as to how this might occur.

In time, the two strongest predictions of the Clonal Selection Theory were verified, namely that antibody pre-exists the exposure to antigen, and that precursor cells only bear antibody receptors of one specificity.

All the originators of the Clonal Selection Theory, except Lederberg, speculated that tolerance was uniquely established during development and perhaps neonatally. Lederberg envisaged the maintenance of tolerance to be a process continuing throughout life in face of the continuous generation of precursor cells. Lederberg's view was more easily reconcilable with the fact that autoimmunity could arise in adults. He envisaged that the critical antibody-bearing cell, when first born, would die if it interacted with antigen, and would in time give rise to an inducible cell only if it had not interacted with antigen, see Figure 9. Thus, precursor cells with receptors able to interact only with foreign antigens would accumulate.

The originators of the Clonal Selection Theory also speculated on the nature of the generation of diversity (GOD) of antibody genes. Some, including Lederberg, speculated that genes coding for antibody chains are not inherited from parents in the usual way. Rather, individuals inherit assembly kits that allow different cells of an individual to generate many different genes by a process involving chance events. They visualized that, in this manner, a limited amount of DNA could encode an enormous number of antibody chains. These speculations on self-nonself discrimination and on GOD inspired much thoughtful experimentation and led to great progress in the field.

Chapter 3

MY INTRODUCTION TO SCIENCE AND TO IMMUNOLOGY, AND THE FORMULATION OF THE TWO SIGNAL MODEL OF LYMPHOCYTE ACTIVATION

Preface

At this point the narrative takes a different turn, as I describe how I became engaged in immunological questions. The first two chapters of this book describe the state of the field before I started my immunological research. I was a witness of and a participant in the evolution of the field from the late 1960s to the present day. I now give the best account I can of both the field and of my involvement.

How I came to be interested in science

I came from a scientific family, my father being a physicist and my mother having studied mathematics as an undergraduate. At about the age of twelve, I became interested in both philosophy and in physics, and was initially torn between these two disciplines. I came to the decision, at about the age of thirteen, that I should try to become a physicist. I was more convinced physics dealt with an approximation of reality, and I therefore imagined it would be more rewarding than studying philosophy. I believe the pursuit of my philosophical interests helped develop my theoretical bent. I was accepted by Cambridge University to study natural sciences when I was sixteen. I studied physics. As I did my undergraduate studies, I came to feel I did not have the gifts to become a theoretical physicist, and was seeking a less demanding field where theoretical considerations could be significant for the field's development.

I was pondering in 1962 on what I might do after I graduated, anticipated to occur in 1964. My brother Mark, three years older than I, had studied chemistry and had started his graduate studies in 1961. He had decided to apply to do gradu-ate studies at the Cambridge Laboratory of Molecular Biology, and was accepted. This laboratory was a Cambridge phenomenon. It can well be argued that it was pivotal in the genesis of the discipline of molecular biology. Several of its scientists had recently received Nobel prizes. Sanger had received one for determining the first amino acid sequence of a protein. Perutz and Kendrew received their prize for determining the first three dimensional structure of a protein, and Crick received his for his contributions to the structure of DNA. At the time, Brenner and Crick were focused on how DNA specifies the nature of proteins, and on the nature of protein synthesis. Brenner and Crick supervised Mark's graduate studies.

Mark asked Crick for advice on whether I might do graduate studies at the same laboratory. Crick thought that many physicists found it difficult to think biologically, and that I should be encouraged to do graduate studies in protein crystallography. This was a good field for a physicist, even for those who might have difficulties in thinking biologically, and so this field was relatively safe. I had a wish to get into genetics, but advice offered by Crick could hardly be ignored. It was arranged that I would do my graduate studies under John Kendrew, assuming I got a good enough undergraduate degree. Although my final exams did not result in my achieving a first class degree, I was accepted as a graduate student.

There is no doubt that my interest in immunology arose from the incredible diversity of interests of people at the Cambridge Laboratory of Molecular Biology. It might be of interest if I briefly explain how my interest in immunology came about.

My interest in immunology

A characteristic of the Laboratory of Molecular Biology was the enormous amount of time spent on scientific discussion. There was morning coffee, lunch, and after-noon tea in the top-floor cafeteria, where people from different disciplines mingled. The second floor held the Division of Molecular Genetics, with Sydney Brenner and Francis Crick sharing a big office. Another ritual, crucial to my development and path in life, were Saturday morning get-togethers of five to seven junior people, post-docs, and graduate students, with Sydney Brenner. Apart from me, all participants were from the second floor. These get-togethers were held in the "kitchens", the rooms where wash-up was done and media made during the week. On Saturdays, large beakers were employed to heat water and make Nescafe. It was not very refined to the taste, but that was not why we gathered. All sorts of science, and some politics, were discussed. I joined this group on first coming to this laboratory, as my brother Mark attended regularly. I continued to attend when Mark left Cambridge to take up a postdoctoral fellowship at Stanford, California, with Paul Berg.

Sydney Brenner started elaborating one Saturday in 1966 on a paper he and Cesar Milstein had just submitted to *Nature*. A major question in immunology at that time was the origin of antibody genes. Their paper proposed a particular mechanism by which the generation of these genes could occur somatically, i.e. in the lifetime of the individual. Their proposal was based upon the knowledge that immunoglobulin chains had an amino-terminal variable domain of about 120 amino acids, joined to one or more constant domains of similar size. Sydney explained their ideas as they would apply to the generation of light chain diversity. They envisaged that one light chain gene was handed down from each parent to their offspring. An enzyme acted on this gene to cause a break in one of the two strands of the DNA at the site corresponding to the join between the variable and constant domains. The broken DNA strand was degraded in the region corresponding to the variable region of the light chain, to varying extents, on different occasions, and then repaired by a sloppy DNA polymerase, introducing mutations into the repaired strand. An expectation based on this model was that the diversity between different light chains would be greatest on the variable side of the junction between the variable and constant region, as this was the region most often sloppily repaired. Variability would be least at the other end of the variable region, the amino-terminal end, as this part was furthest from the initial break in the DNA, and the region whose corresponding DNA was least often repaired.[35]

I thought about the ideas Sydney had espoused over the next week. There was something about this proposal that troubled me. The next Saturday, I tried to articulate my concerns to him. I thought that a highly variable sequence over a considerable length of the light chain would have a highly variable three-dimensional shape. Even the variable parts of antibodies had to have some common overall structure, I thought. There might therefore be a need for non-variable amino acid residues at certain positions or regions in the variable domain. I tried, in a moment of enthusiasm, to make my point by saying you could not make antibodies out of scrambled eggs. Scrambling eggs denatures the proteins of the egg, which consequently lose their normal structure and become insoluble, and so come out of solution. Sydney was marvelous. He took my thoughts seriously, and said he would ask Cesar whether there was any pattern of invariant residues in the variable region. Cesar re-examined his data from this perspective and thought there might well be such invariant residues, with the limited observations then available. The validity of the idea of invariant amino acid residues in the variable region became clearer with further studies. This conclusion made the Brenner/Milstein model less plausible.

This adventure spurned my interest in immunology. I started trying to think about different aspects of the subject. I talked to Cesar Milstein, and he suggested which book I should read as an introduction to the field. I started it, but did not read it through because too many of the proposals seemed wild. I was ignorant and unaware that the author, Burnet, was a giant of the field. Nevertheless, this book made me aware of the immune system's attribute of self-nonself discrimination, and of an experimental generalization. Burnet made the comment that small molecules were not immunogenic, meaning immunizing with them would not result in the production of antibody. Immunogenic molecules were always macromolecules, i.e. relatively large. I wondered why there was this requirement on a molecule to be large for it to be able to induce the formation of antibody. Relatively small molecules could bind to antibody perfectly well. The logical inference would be that the binding of a small molecule to the antibody receptor of a precursor cell was insufficient to induce the precursor cell to multiply and its descendants to produce antibody. I lodged in my brain the idea that molecules had to be big to be immunogenic, with the feeling that this might provide the clue to something.

An accident and its consequences: formulation of the one cell/multiple cell model for the inactivation and activation of precursor cells

One of the projects that constituted my real research, directed at obtaining a PhD thesis, involved the use of a small molecule highly reactive with proteins. I took care to wear gloves in handling it. One night, as I fell asleep, I felt a throbbing in the tip of a finger. I had used the chemical that day. In the morning, I woke up to find that my fingertip was white and without any feeling. I guessed that the glove had leaked. I went that morning to the hospital, explaining what I thought had happened. I could see the medical personnel were somewhat skeptical. However, they came around to my view when they drove a needle into the numb fingertip and I did not react.

During the next few days, a sharp line developed between the white and dead part of my fingertip and the rest of it. I imagined that my immune system was reacting to my altered proteins as foreign, and leaving those not damaged alone, resulting in this sharp demarcation between the damaged and healthy tissue. This discrimination between self and foreign (nonself) antigens was one of the recognized attributes of the immune system. How self-nonself discrimination is achieved was, and is, a fundamental question of immunology. My intuition was that this problem would most probably have a simple and elegant solution. After all, there was nothing intrinsic that distinguished all self from all foreign antigens. I reasoned by myself or, more likely, I had been convinced by reading the ideas of others, that the immune system must distinguish self from foreign antigens on the basis that self antigens, in contrast to foreign antigens, appear early in development and are then continuously present. I also knew that small, univalent molecules could bind antibody, but were unable to cause their precipitation, as this required that the antigen be large enough to bind more than one antibody molecule to form sizable lattices.

From my scattered reading, I had also gathered the idea that precursor cells, as envisaged in the Clonal Selection Theory, could interact with antigen to be either activated or inactivated. This perception turns out to be most probably correct, but was different from what Lederberg had postulated. Activation of precursor cells led to their multiplication and the differentiation of their descendants to produce antibody. Inactivation resulted in their becoming insusceptible to activation signals, perhaps through their death. It appeared that there must therefore be two distinct ways by which antigen can interact with precursor cells, with different consequences.

One day in 1967, during my graduate studies, it dawned on me that a couple of ideas could explain a fact or two, and give a potential insight into how antigen can interact in one way with precursor cells to inactivate them, and in another way to activate them. Moreover, my ideas also explained how self-nonself discrimination might be achieved. Actually, I am pretty sure that the need to explain self-nonself discrimination led to the ideas on how antigen activates and inactivates precursor cells. I cannot really know as much of this "work" was done subconsciously. The result of these unconscious struggles became apparent as I became conscious of connections between ideas, rather as a butterfly breaks out from its chrysalis.

I envisaged that a single precursor cell, upon interacting with antigen, could only be inactivated. In contrast, I envisaged that the activation of this precursor cell required the antigen-mediated interaction of this cell with other antigen-specific cells. These proposals were appealing for two main reasons. First, small, monovalent molecules can bind to antibody but do not precipitate with them, in contrast to larger, multivalent antigen molecules. Perhaps similarly, small, univalent molecules are also not immunogenic, only larger, foreign molecules. The greater the number of sites on an antigen recognizable by antibody, the more precursor cells there would be specific for it, and the easier it would be to achieve the antigen-mediated interaction between the precursor cell and the other specific cells, that I envisaged to be required for precursor cell activation. The proposed rules governing the inactivation and activation of precursor cells explained the observation that large but not small foreign molecules were immunogenic. This generalization was one I had gleaned from reading Burnet.

More excitement arose from my realization that these proposals could account for self-nonself discrimination. People envisaged that precursor cells were first generated at some time in development and were continuously generated thereafter. Consider a self antigen present at the time the first precursor cells are generated, see Figure 10. The first precursor cell generated that is specific for this self antigen could not be activated, as there are no other antigen-specific cells with which it could interact. This precursor cell would therefore be inactivated by the self antigen. Precursor cells specific for this self antigen would be inactivated as they were generated, one or a few at a time, by virtue of the continuous presence of the self antigen. Consider what happens to the precursor cells specific for a foreign antigen. The foreign antigen is not present when precursor cells are first generated. Those precursor cells generated and specific for the foreign antigen will accumulate in its absence. When the foreign antigen later impinges upon the immune system, a population of precursor cells specific for this antigen will be present. The foreign antigen can then mediate the interaction between the specific cells required for their activation. This accounting of how self-nonself discrimination might be achieved is depicted in Figure 10.

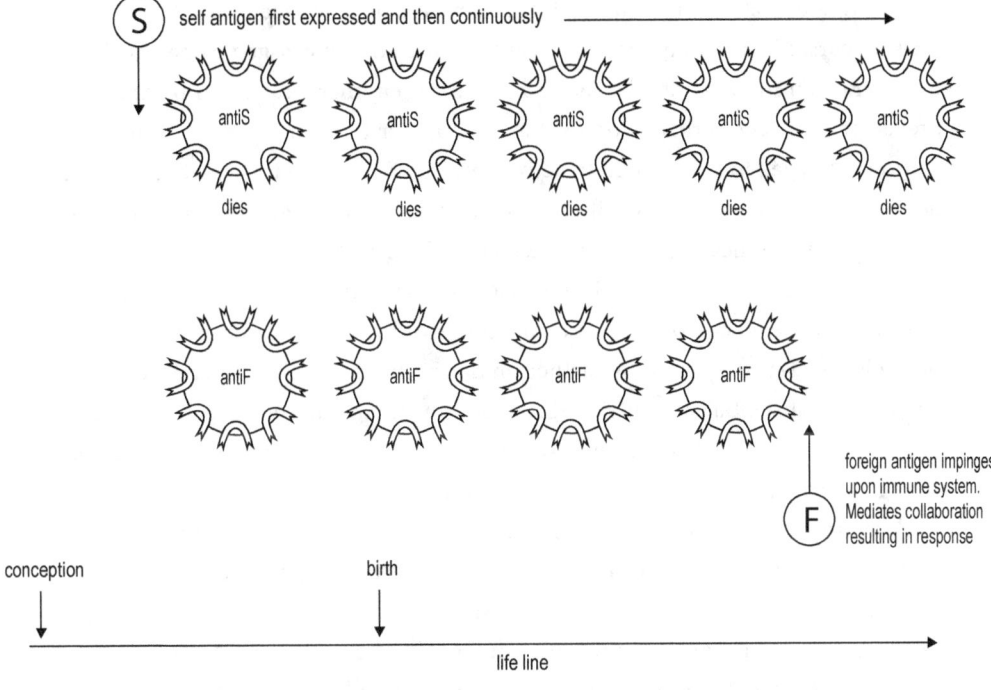

Fig. 10. Illustration of how the requirement for the antigen-mediated cooperation of lymphocytes for their activation, but not for their inactivation, accounts for self-nonself discrimination

A chance finding of a paper provides support for my proposals

As a graduate student still in Cambridge, I randomly dipped into journals that published immunological papers in the hope of stumbling on something pertinent to those immunological ideas I was particularly excited about. I found a recent paper[36] that reported on the requirements to obtain an antibody response to a very simple macromolecule, *poly-L-lysine (PLL)*, a synthetic polypeptide chain consisting of a string of lysines, a common amino acid. One inbred strain of guinea pigs could produce antibody when immunized with PLL, whereas another strain could not. These guinea pigs were respectively called responders and non-responders. The paper showed that non-responders would produce anti-PLL antibody when immunized with PLL conjugated to the foreign and immunogenic protein, **bovine serum albumin**, or **BSA**. In other words, PLL was a macromolecular hapten.

One had to infer from the production of anti-PLL antibody in non-responder guinea pigs, when immunized with PLL-BSA, that there were precursor cells with antibody receptors able to bind PLL in these guinea pigs. Why was PLL-BSA but not PLL able to induce anti-PLL antibody in these guinea pigs? BSA itself was immunogenic and so, according to the proposal that activation required the antigen-mediated cooperation of lymphocytes, there must be a sufficient number of anti-BSA precursor cells to support the cellular cooperation envisaged to be necessary to generate anti-BSA antibody. As we have just seen, there must also be anti-PLL precursor cells, but perhaps not a sufficient number to support the cellular cooperation envisaged to be required to activate precursor cells. In this case, PLL would not be immunogenic. The anti-PLL precursor cells are induced, however, when the non-responder guinea pigs are immunized with PLL-BSA, presumably involving an interaction between the anti-PLL precursor cells and the more prevalent anti-BSA cells, as a consequence of their both binding to PLL-BSA. I came to realize later that this idea was supported by further findings.[37] It was possible to give new-born guinea pigs substantial amounts of BSA, rendering them unresponsive or tolerant to BSA as young adults. This neonatal administration of BSA presumably depleted the young guinea pigs of BSA-specific precursor cells.

A report described that when such guinea pigs, unable to respond to BSA, were challenged with PLL-BSA, PLL-specific antibody was not produced. This led me to suggest that the conjugate PLL-BSA was able to activate anti-PLL precursor cells when given to normal guinea pigs due to the presence of BSA-specific cells, a situation that would not occur in guinea pigs made tolerant to BSA. This seemed a plausible explanation of the observations to me, but the authors of the paper had appeared to be at a loss as to how these observations might be interpreted. I also came across a series of papers on "breaking the unresponsive state", which were supportive. I describe them below, under "Examples of carrier effects."

Advice and help from on high

I was very excited by these ideas. They are conveniently referred to as the one cell/ multiple cell model for the inactivation/activation of precursor cells. I discussed them with various people, including Crick.

Crick had pinned on a notice board, beside his office, a card that included something along the lines: "For scientific discussion, please enter." I availed myself of this invitation to discuss both my PhD thesis work and my ideas on immunology. I had discussed with him some of my earlier immunological ideas before I had formulated the ones outlined above.

I started getting advice from Crick, well before the end of my graduate studies, about whom I might do my postdoctoral studies with. He initially suggested Avrion Mitchison, who worked at Mill Hill, close to London. Av, as he was known to all, was clearly a leading immunologist, and would soon come to the conclusion that the activation of precursor cells is facilitated by the antigen mediated cooperation between this precursor cell and other antigen-specific lymphocytes. I decided to visit him. He said his lab was full but, if I really wanted to come, he might be able to squeeze me in. At the time of this visit, I had not formulated my proposals on the requirements to activate and inactivate precursor cells, and neither had Av come to his view that cell cooperation could facilitate the activation of antibody precursor cells.

Later, as I interacted with Francis more, he realized that I had an unusual love for discussing scientific ideas in a rather theoretical manner. He changed his advice. Crick was a Non-Resident Fellow of the Salk Institute, situated in southern California, and visited there for a few weeks every year. There he got to know Melvin Cohn. Mel was an immunologist who also greatly enjoyed discussing scientific ideas. Francis thought Mel and I would be a good match. Needless to say, I took his advice.

Not surprisingly, Mel accepted me on the recommendation of Francis. I therefore found myself at the Salk Institute for Biological Studies in late July or August, 1968. I had applied for and been awarded a Damon Runyon Postdoctoral Fellowship. I was regarded as something of a curiosity when I arrived as a twenty-five-year-old in Southern California. Few of my age had a PhD and few were unable to drive an automobile. I was unusual in both respects.

The first few moments and months in Mel's laboratory

I realized within minutes of meeting Mel, despite my jetlag, that his personality had something in common with Sydney Brenner's. They both were theatrical and thoroughly enjoyed performing.

I was intent to immerse myself in the immunological literature pertinent to my ideas. I wanted to calm myself and take careful stock. Mel at the time was most interested in the origin of the genes coding for the diversity of light and heavy chain immunoglobulin genes. Indeed, I believe it was Mel who first referred to the problem of how these genes arose as the problem of the *generation of diversity*, or as the problem of the GOD.

However, my idea of immersing myself in the literature, and of slowly developing my ideas into a form where they were fit for public display and discussion, was not to be. Mel had been invited to a small meeting on immunological tolerance, only for invited participants, held at Brook Lodge, Michigan, from September 18-20, 1968. Moreover, he was asked to summarize the meeting. It was limited to about forty people, including most of the significant researchers of the field, including Baruj Benaceraff, Av Mitchison, Klaus Rajewsky, and Bill Weigle. It was only natural that Mel tried to bring together the contemporary observations, outlined at the meeting, in terms of the ideas we were in the midst of trying to develop. On returning to the Salk Institute, Mel gave me the transcript of the meeting. I was somewhat upset. I felt Mel had shown he did not know some of the critical experiments that supported our proposals. Moreover, the meeting was to be published in book form.[38] In the circumstances, we decided to quickly write an account of the ideas as they then stood and submit them as a paper to *Nature*. The journal received our article on October 17, less than a month after the Brook Lodge Meeting, and accepted it the next day; the proofs were corrected over the phone and our article was published in the November 2, 1968, issue,[39] about two or three months after my arrival in Mel's lab.

The style of science and of life at the Salk Institute and in Mel's lab

The Salk Institute was somewhat different from the MRC, as I shall indicate later, but the two institutions had two features in common. I would argue these features were central to the particular pleasures of working there and allowed individual researchers to be exceptionally fruitful. Many of the researchers at both places attempted to answer the most fundamental questions they could envisage. Leslie Orgel was a Resident Fellow at the Salk Institute, had been a theoretical chemist in Cambridge, and had interacted and collaborated with Crick on the chemical nature of mutations. He came to the Salk Institute with a new initiative: to explore at the chemical level how the origin of life might have occurred. Mel Cohn was involved in trying to understand the origin of the genes that code for antibody chains. Mel had a substantial grant to generate myeloma tumors in mice—that is, tumors of antibody producing cells—and to characterize the myeloma proteins they produced. These proteins were believed, most probably correctly, to reflect the nature of normal antibody molecules. The availability of large amounts of purified myeloma proteins allowed their chemical nature to be analyzed. For example, the amino acid sequence of their light chains could be determined. This work in Mel's lab led to insights into how the genes coding for immunoglobulin chains might be generated.

Renato Dulbecco was also a Resident Fellow and interested in how viruses interact with host cells to transform them into cancer cells, work that led to his Nobel Prize. A recent arrival was Robert Holley, who later received the Nobel Prize for his work on the structure of an important class of molecules involved in protein synthesis, namely transfer RNA. Roger Guillemin, the neuroendocrinologist, came later, and also received the Nobel Prize for his contributions. I give this record as a testament to the substantial significance of the research carried out at the Salk Institute and, more importantly, I would suggest, as a testament to a way of doing science that nurtures such achievements.

The second feature, common to both institutions at the times I was there, was the lack of pressure to be productive in the short term. This largesse provided the individual with the opportunity to explore and develop new interests. It is to Mel's credit that this was particularly apparent in his lab. There was also an inevitable downside. When I came to the laboratory, I don't think anyone there had a strong background in immunology, though it could be argued that Mel was the exception. This made it difficult to get experiments off the ground, as good, practical advice was not close at hand. For example, most immunologists at that time knew how to inject mice intravenously, but I think no one in Mel's lab had acquired this art when I arrived.

The state of some immunological questions at the time I became a postdoctoral fellow

A major meeting devoted to immunology was held in 1967 at the Cold Spring Harbor Laboratories (CSHL) in the Eastern United States, where participants described their latest experiments and expressed their current views. I arrived in Mel Cohn's laboratory in the autumn of 1968. As a graduate student at Cambridge, I had studied a comprehensive and interesting review article by Dresser and Mitchison,[40] published in the 1968 volume of *Advances in Immunology*. This article was somewhat out of date due to the greater time lag between submission and publication in those days. However, while at Cambridge, I was not aware of the proceedings of the CSH meeting. I started reading them[41] voraciously on my arrival in Mel's lab, as well as other recent books and journal articles. I talked with Mel almost every day for at least an hour.

I was initially disappointed to find some ideas I had formulated, and considered to be novel, had recently been espoused by others. For example, some immunologists, particularly Av Mitchison[42] and Klaus Rajewsky,[43] suggested, in their contributions to the Cold Spring Harbor Proceedings that the activation of an antibody precursor cell was facilitated by its antigen-mediated interaction with other specific lymphocytes. This proposal was expounded upon by Niels Jerne in his summary of the meeting. Nevertheless, I also found that their views differed substantially and critically from the proposals I had thought of. I think it worthwhile to discuss these differences, not only for their intrinsic interest, but because the considerations involved still bear critically on current and central issues, almost fifty years later.

Examples of carrier effects

Most of the intellectual energy of immunologists during the 1950s and early 1960s had been directed at issues related to the formulation of the Clonal Selection Theory. As we have seen, Lederberg proposed in his 1959 *Science* paper that precursor cells can be inactivated by antigen when they first arise. If not inactivated, they in time differentiate into an inducible state. This scheme was insightful, but is not referred to or discussed in Dresser and Mitchison's 1968 article in *Advances in Immunology*.[40] The discussion they presented took place in a different framework. I shall first give an account as to how I think this framework arose.

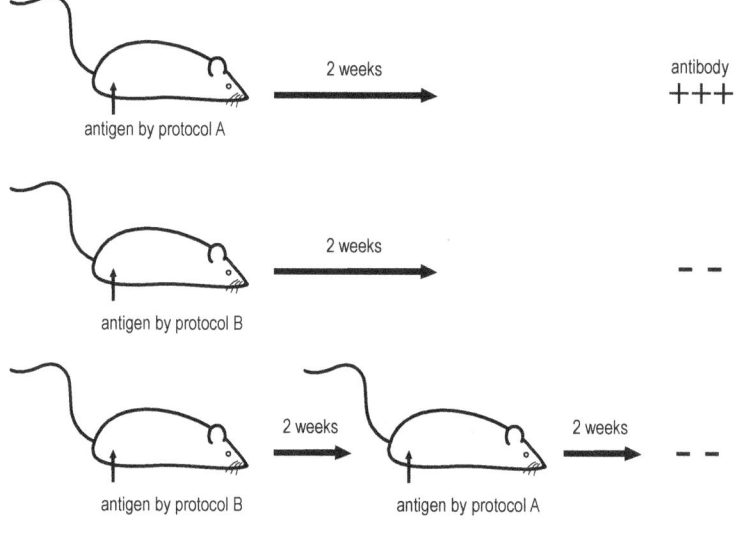

Fig. 11. Schematic protocol for demonstrating the establishment of unresponsiveness in immunocompetent mice and other animals

Diverse observations had decisively shown that the sustained exposure of developing animals to a foreign antigen led to the immune system regarding the antigen as self, as epitomized by the experiments of Medawar and colleagues in neonatal mice[27] and, I would suggest, even more clearly by the experiments of Hasek and his colleagues in developing chicks.[44] Immunologists subsequently found that the administration to many young animals—mice, rats, chickens, and rabbits—of some antigens were, under some conditions, immunogenic and, under other conditions, were not. Most interestingly, a common finding was that giving the antigen under non-immunogenic conditions often made the animals unresponsive to a subsequent challenge that led to an antibody response in naïve animals, see Figure 11. This body of observations led to the belief that the processes underlying the establishment of tolerance were not limited to the developing animal. Dresser and Mitchison's article probably reflected a widely held belief: a precursor cell could, on interaction with antigen, be either activated or inactivated. In addition, most envisaged that the simple binding of antigen to the antibody receptors of the precursor cell led to its activation, i.e. to its multiplication and the differentiation of its descendants into antibody-producing cells. We refer to this model as the **Simple Antigen Binding Model** for precursor cell activation. However, observations had accumulated that could not be accounted for by this model. It is helpful to discuss three of these classical studies to make vivid the issues they raised and the insights and perceptions they led to. Moreover, some of these studies appear to shed light on the circumstances leading to autoimmunity, as we shall see and, perhaps surprisingly, are also pertinent to current debate.

Experiment 1

The first set of experiments, reported in 1961, involved giving a group of newborn rabbits a massive dose of a prominent bovine serum protein called **bovine serum albumin,** or **BSA**. Naive rabbits, at three months of age, can respond well to this foreign antigen, but those exposed as neonates to massive amounts of it do not, as shown in groups 1 and 2 of Figure 12. Immunologists imagined that this unresponsiveness to the BSA challenge was equivalent to the natural unresponsiveness or tolerance to self antigens. This seems a plausible inference. Naïve, three month-old rabbits can also respond to **human serum albumin, HSA,** the protein in human serum homologous to BSA. Moreover, about 10% of the anti-HSA antibody produced in naïve rabbits reacts with BSA (see group 3). The antigens BSA and HSA significantly *crossreact*. The most intriguing observation is that rabbits given a massive dose of BSA at birth and challenged at three months with HSA, produce anti-HSA antibody, and that some of this antibody reacts with BSA, i.e. anti-BSA antibody is produced (group 4).[45] How can we explain this?

To repeat, anti-BSA antibody is produced on challenging BSA-unresponsive rabbits at three months of age with HSA (groups 3 and 4). This observation must mean, in the context of the Clonal Selection Theory, that there are precursor cells with receptors able to bind to both BSA and HSA. Why are these anti-BSA/HSA precursor cells activated by HSA but not by BSA? It seems likely that this is because there are more HSA-specific than BSA-specific cells to support the interaction of the anti-BSA/HSA precursor cell with other specific cells. I had surmised that such an interaction is required for the activation of precursor cells.

These observations, when first reported in 1961, appeared to be enigmatic. When I first became aware of them in 1967, I could find no explanation of them in the literature. However, I felt they could be naturally explained if the activation of precursor cells required cellular cooperation.

Another observation made in this system was important. It was found that, if the BSA-unresponsive rabbits are challenged at three months with HSA and simultaneously given sufficient BSA, no antibody able to bind BSA is produced, though antibody able to uniquely bind to HSA is still generated, see group 5 of Figure 12. This type of observation, repeated in different systems, led to the suggestion that the precursor cells, able to be activated by antigen to generate descendants that produce antibody, can also interact with antigen in a manner that inactivates the precursor cell. This type of observation supported the idea that there are two ways antigen can interact with precursor cells, one leading to its activation, the other to its inactivation, and that there was competition at the level of the precursor cell between these two processes. In the context of the experiment just described, HSA was *immunogenic* and BSA was **tolerogenic.**

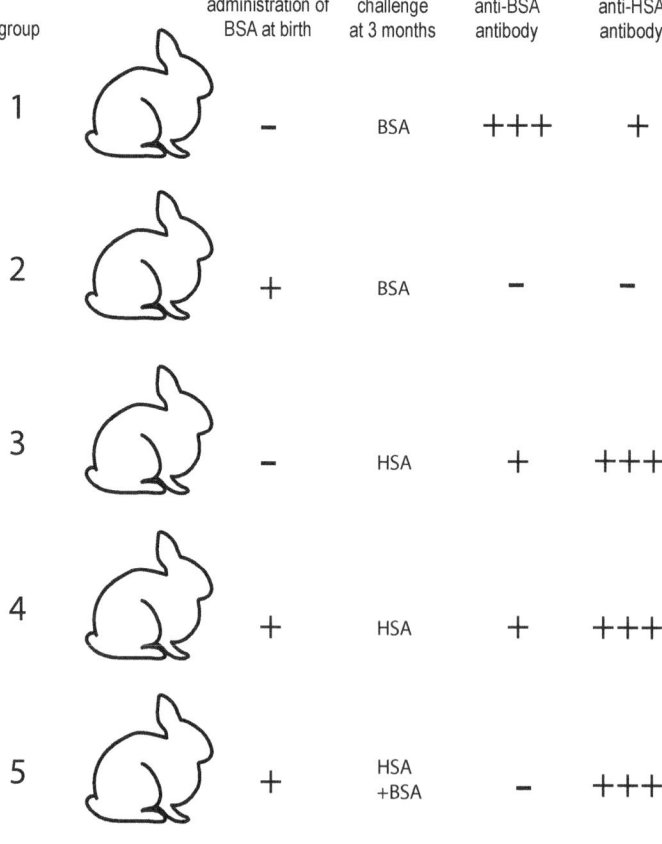

group		administration of BSA at birth	challenge at 3 months	anti-BSA antibody	anti-HSA antibody
1		–	BSA	+++	+
2		+	BSA	–	–
3		–	HSA	+	+++
4		+	HSA	+	+++
5		+	HSA +BSA	–	+++

Fig. 12. Breaking the unresponsive state against BSA by a challenge with the crossreacting antigen HSA.

After discussing the other two experiments, we shall consider what generalizations might be drawn, and discuss how this experimental model most probably provides an insight into some circumstances under which autoimmunity is generated.

Experiment 2

The second study again employed rabbits and three related porcine enzymes, each containing four subunits. Two of the enzymes have four identical subunits, and are referred to as A_4 and B_4, and the other enzyme contains two A and two B subunits, and is referred to as A_2B_2.

Rajewsky found a strain of rabbits that makes robust antibody responses when challenged with A_4 but not when challenged with B_4. However, immunization with A_2B_2 resulted in strong anti-A and anti-B antibody responses as assessed by their reactivity with the A_4 and B_4 antigens.[43] Therefore, according to Clonal Selection Theory, there must be B-specific precursor cells present in these rabbits. Why were these B-specific precursor cells inducible by A_2B_2 but not by B_4? The same pattern suggests itself! These scarce anti-B precursor cells need anti-A cells to be activated. The anti-A precursor cells are more prevalent than the anti-B variety.

Another observation made in this system parallels observations made in other systems. This type of observation appears to have been pivotal in people's thinking. It was found that B_4 unresponsive rabbits, given a challenge of B_4, could subsequently produce anti-B_4 antibody when immunized a few weeks later with A_2B_2. This was interpreted to mean that, in addition to B_4 being non-immunogenic, it was also unable to inactivate precursor cells, and so was ***non-tolerogenic***. We shall shortly return to this type of observation and its interpretation.

Experiment 3

It was found in a number of systems that the degree of aggregation of an antigen had a profound impact on its immunogenicity. For example, commercially available BSA is not homogenous. The BSA protein has a molecular weight of around 70,000 daltons, but commercial BSA has a substantial amount of higher molecular weight material, most of which is aggregated BSA. Removal of this aggregated material, leading to a preparation called deaggregated BSA, results in a loss of the immunogenicity of the preparation. Moreover, the administration of this deaggregated BSA to guinea pigs results in an unresponsive state; guinea pigs exposed to deaggregated BSA do not produce anti-BSA antibody on a subsequent challenge with commercial BSA that results in an antibody response in naïve guinea pigs.[46] This type of experiment is again most readily understood if there is competition between the activation and inactivation of precursor cells, and if aggregated BSA is *immunogenic* and deaggregated BSA is *tolerogenic*.

A potential framework to provide a coherent explanation of carrier effects

I shall lay out the issues, concerning the activation and inactivation of antigen-specific lymphocytes, that I felt in the late 1960s had to be faced. I explain why I believe our model resolved them. I think it best to make certain simplifications in order to best display the essence of the argument. We shall later revisit these simplifications and come to the conclusion that they do not affect the validity of the argument. Abandoning them will lead to further questions, which will be discussed later.

The simplifications I shall make can be collectively expressed by stating that the interaction of antigen with a single precursor cell results in its inactivation, and that its activation requires the antigen-mediated interaction of this precursor cell with other, similar precursor cells specific for the same antigen. I now consider how the observartions made in these three experimental systems can be accounted for in terms of the one cell/multiple cell model for the inactivation/activation of precursor cells.

There must be quantitative considerations when thinking about the consequences of these proposed rules. It is clear that a sufficient amount of antigen will inactivate a sole precursor cell. How many precursor cells are needed to allow the antigen-mediated interaction required for their activation? We do not know. However, let us imagine it is one hundred for the purpose of exploring certain situations.

Experiment 3 shows that deaggregated BSA inactivates the BSA-specific precursor cells. In practice, if the guinea pigs are not further exposed to BSA to maintain the unresponsive state, they regain their ability to respond to a BSA challenge in about a hundred days. These figures suggest that new anti-BSA precursor cells are generated at about one per day.

Consider now experiment 2. In this case, the administration of B_4 to B_4-unresponsive rabbits does not dramatically affect the production of anti-B antibody, following a challenge with A_2B_2 a few weeks later. Suppose the administration of B_4 *does* inactivate anti-B precursor cells, just as the administration of non-immunogenic, deaggregated BSA inactivates BSA-specific precursor cells. As A_4 is immunogenic, there must be more than a hundred anti-A precursor cells, if that is the minimum number needed to initiate this response. Thus, a challenge of a rabbit with A_2B_2 that has just one anti-B precursor cell will result in the activation of this anti-B precursor cell and the production of anti-B antibody, even if not that much antibody is produced. Challenging with A_2B_2 a few weeks after administering B_4 may well result in substantial anti-B_4 antibody, according to the simple ideas on the activation and inactivation of precursor cells that we are exploring, so long as a few new anti-B precursor cells have been generated.

Several experiments of this type in other experimental systems showed that pre-exposure of an animal to a non-immunogenic molecule N did not significantly affect the anti-N antibody response obtained when the animal was subsequently challenged *some time later* with N-C, where C is an immunogenic carrier. This gave rise to the proposition that *non-immunogenic molecules are also non-tolerogenic*, a mantra of the time. This generalization, predominant at the time I joined Mel's laboratory in California in the autumn of 1968, was contrary to the ideas I had developed whilst a graduate student. Moreover, this generalization was also contrary to other findings. It was inconsistent with the observation that deaggregated BSA was both non-immunogenic and tolerogenic, as seen in Experiment 3 above. Also, it was inconsistent with the observation that the administration of BSA could prevent the production of anti-BSA antibody in BSA-unresponsive rabbits normally obtained on challenging them at three months of age with HSA, see group 5 of Figure 12. In both these cases, the administration of non-immunogenic molecules could block the production of antibody specific for those molecules. In addition, and at a more general level, it was difficult to see how self antigens, presumably not immunogenic, could inactivate precursor cells, if this proposition was really correct.

When we published our model, within about three or four months of my joining Mel's laboratory, I received a personal letter from Av Mitchison. He naturally used the terminology of the day. He said, and this is an accurate quotation, "I think you deeply cloud the issue by using the carrier effect to account for the immunity/tolerance decision. I don't at all see why one cannot have both all sorts of lovely carrier effects and also something else, e.g. non-specific stimulation from macrophages." It seemed clear he understood our view and disagreed.

Refining our ideas on the activation and inactivation of lymphocytes

I soon recognized that I was uneasy about some aspects of the proposals made in the 1968 *Nature* paper, about which, to be honest, I was somewhat embarrassed. I will not digress to describe these here, but will rather give an account of how the ideas further evolved in a constructive way.[47] Before doing so, I should address some other significant aspects of our paper that reflect considerations I have so far avoided, by virtue of discussing the essence of the ideas in terms of the simplest model conceivable. This does not mean, however, that the considerations are unimportant.

We recognized, as did all immunologists, that there was a precursor cell for the antibody producing cell which, for convenience, I will refer to as the **antibody precursor cell**. The question that naturally arises in the context of the proposed interaction of this antibody precursor cell with other specific cells, and envisaged to be required for the antibody precursor cell's activation, is whether the antigen-specific cooperating cell is of identical type to the antibody precursor cell. It was known at the time that there were different classes of antibody with different functions. We suggested that the cooperating cell produced a special class of antibody that we initially dubbed as "carrier-antibody". Thus, we envisaged that the cooperating cell was a different type from the typical antibody precursor cell. This turned out to be a correct guess.

We can approach the second consideration incisively if we define some terms to express what others had already surmised concerning the production of antibody. The antibody precursor cell does not secrete antibody but bears antibody receptors on its surface. When this antibody precursor cell is activated, it divides and some of its progeny give rise to *antibody effector cells* that produce and secrete large amounts of antibody. We supposed that this generation of *effector cells*, by activating precursor cells, might represent a general process, involving, for example, the proliferation of these precursor cells and the differentiation of their descendants into effector cells. The second question was whether the cooperating cell, able to efficiently interact with the antibody precursor cell and resulting in its activation, is really a precursor cell, or whether it is an effector cell, being generated when antigen activates a carrier-antibody precursor cell.

This question is important but not essential. It is not essential in the sense that cell cooperation is required for antibody precursor cell activation according to both alternatives, and so both alternatives are compatible with the one cell/multiple cell model for lymphocyte inactivation/activation.

We suggested that carrier antibody effector cells are much more efficient cooperators in activating antibody precursor cells than are carrier antibody precursor cells. This possibility allows the process of the generation of efficient cooperators to be more readily the subject of regulation. We shall see that we were lucky in this choice, as it turns out to be correct. However, this alternative leads to a further question: how are carrier antibody precursor cells themselves activated? We envisaged that this process would optimally require the antigen-mediated interaction between the carrier antibody precursor cell and a carrier antibody effector cell. I personally still believe this proposal to be essentially correct, as will become apparent later.

This proposal raises yet a further and obvious question. Our conceptual scheme postulates that carrier antibody effector cells are themselves required to optimally activate carrier antibody precursor cells to generate more carrier antibody effector cells; carrier antibody effectors cells are required to generate more of them. Where could the first carrier antibody effector cell come from? This question is referred to as the *priming problem,*[48] and is central to contemporary discussion.

Analogous priming problems often occur in biology. We need life to beget life, so how did life originate? Ribosomes are particles where proteins are synthesized, and proteins are major components of ribosomes, so we need ribosomes to make ribosomal proteins! Where did the first ribosomes come from? The simplest solution to the priming problem, in the context of carrier antibody, is that, although carrier antibody effector cells are the most efficient cooperating cells, the interaction of carrier antibody precursor cells can lead to their mutual activation, albeit less efficiently than if carrier antibody effector cells are involved in the interaction.[48] We will return to this as we consider more contemporary ideas on how immune responses are initiated.

My first wish on arriving in Mel's laboratory was to focus on whether the literature was in accord with my model of one cell/multiple cells for lymphocyte inactivation/activation. I had initially paid some attention, though not sufficiently focused, on how the antibody precursor cell and the cooperating cell might interact. The nature of this interaction became the subject of much discussion between Mel and me during the following years, and of considerable controversy in the literature. However, this general focus could only arise once there was broad agreement in the immunological community that the activation of antibody precursor cells really does require the antigen-mediated collaboration between antigen-specific lymphocytes.

Cellular cooperation is usually required to generate an antibody response

A series of papers soon led to a consensus that the induction of antibody required cellular collaboration. A set of papers came out around the same month as our *Nature* paper. These papers, by Jacques Miller and some Australian colleagues, elaborated upon a finding by Henry Claman of the United States. Claman had discovered that lethally irradiated mice, when reconstituted with either bone marrow cells or *thymocytes*, cells from the *thymus gland*, and challenged with antigen, could not make antibody to the antigen. However, when reconstituted with both bone marrow cells and thymocytes and challenged with antigen, they made robust antibody responses.[49] These observations led to the suggestion that both bone marrow cells and thymocytes are required to generate an antibody response. Miller and his colleagues provided evidence that the antibody producing cells were descended from the bone marrow cells.[50] Bone marrow cells thus contained the antibody precursor cell. What was the role of the thymocytes? We shall shortly return to this critical question.

Jacques Miller's role in these studies, following up on Claman's discovery, was not accidental. He had already shown that surgical removal of the thymus gland, in the chest cavity and just below the top of the rib cage, from one day-old mice, led to immunoincompetence of these mice as adults. It was also known that mice, given a normally lethal dose of irradiation, would survive if given bone marrow cells that contain *stem cells*. These cells multiply and differentiate into many different cell types, and also give rise to more stem cells. These findings may have collectively inspired Claman to do the experiments showing that both bone marrow and thymocytes, the bulk of the cells in the thymus, were required to obtain a robust antibody response.

Whilst these studies by Miller and colleagues were occurring in Australia, Av Mitchison's group in England was exploiting earlier findings that Av had made.

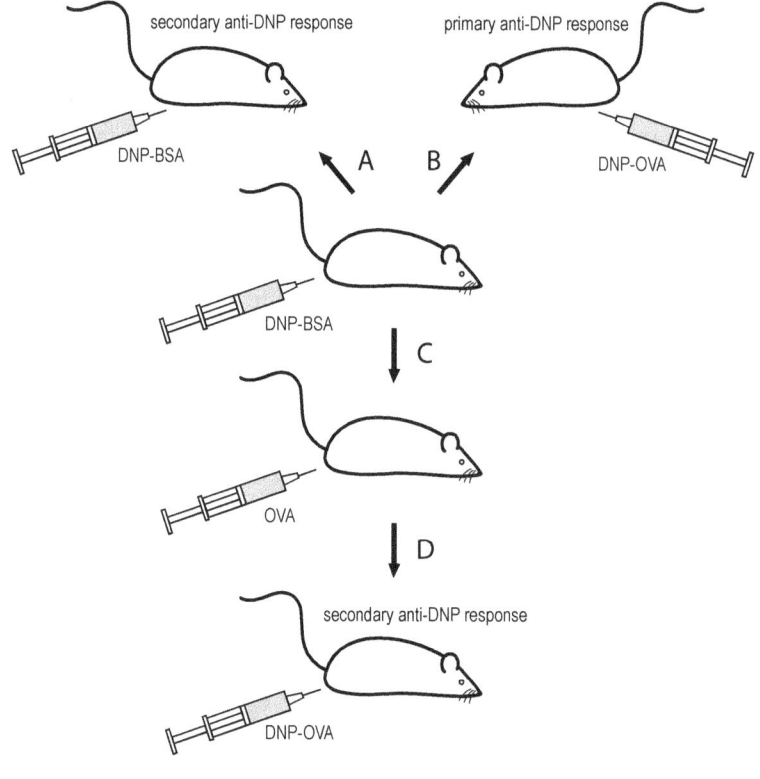

Fig. 13. Priming against both hapten and carrier is required to obtain a secondary anti-hapten antibody response

Mitchison employed hapten carrier systems to explore the requirements to generate secondary antibody responses. We have seen that small, foreign molecules are not immunogenic by themselves, but that antibody can be raised against them when conjugated to an immunogenic carrier. Mitchison did an experiment in which he primed mice with the hapten **dinitrophenyl (DNP)**, coupled to the carrier BSA, i.e. with DNP-BSA, and obtained a secondary response to the hapten on a subsequent challenge with DNP-BSA. In other words, the anti-DNP antibody response was much more rapid and of greater intensity than in a primary response, see A of Figure 13. It was also known that the anti-DNP antibody produced binds well to DNP when the DNP is coupled to another protein, ovalbumin (OVA) for instance.

It was known in addition that BSA and OVA do not crossreact. Thus, a model in which the activation of an antibody precursor cell only requires antigen to interact with its antibody receptor would predict that both DNP-BSA and DNP-OVA could activate the same anti-DNP precursor cell equally well, simply by binding to its receptor. Mitchison showed that priming with DNP-BSA, and subsequently challenging with DNP-OVA, did not result in a secondary anti-DNP antibody response, see B of Figure 13. Somehow, the immune system of the primed mouse "knew" that it had been primed with DNP-BSA.

Mitchison greatly clarified what was going on in his next set of experiments. He primed mice with both DNP-BSA and OVA, and then subsequently challenged them with DNP-OVA, now obtaining a secondary antibody response to DNP, as shown in C and D of Figure 13. This showed that priming against both the hapten and against the carrier is required to generate an optimal secondary antibody response to the hapten. These were some of the critical experiments that led Av Mitchison, in his 1967 Cold Spring Harbor paper, to propose that the optimal activation of a hapten-specific memory precursor cell requires the antigen-mediated interaction of this cell with carrier specific memory cells.[42]

The Great Synthesis

We have seen that neonatal thymectomy, performed on the first day after birth, renders mice unable as adults to respond to most antigens. Neonatal thymectomy is also associated with a deficiency in the spleen and lymph nodes of cells that bear on their surface the Thy1 antigen, recognizable by anti-Thy1 antibody. It turns out that most of the cells in the thymus bear this Thy-1 antigen. It was therefore natural to propose that the deficiency in immune responsiveness of neonatally thymectomised mice was causally associated with the deficiency in Thy1+ cells in the spleen and lymph nodes. This supposition was supported by giving neonatally thymectomised mice Thy1+ cells from a normal mouse, and examining whether this reconstitution now allowed the neonatally thymectomised mice to generate antibody responses. These reconstituted mice could make good antibody responses. These experiments collectively show that Thy1+ cells are thymus-dependent cells, and that they are required to generate robust antibody responses. The Thy1+, thymus-dependent cells became known as *T cells*.

Mitchison took his analysis further. It was known that irradiated mice were immunoincompetent unless reconstituted with immunocompetent cells, such as spleen cells. It was also known that primed mice could mount secondary responses, and that the transfer of primed spleen cells to an irradiated mouse allowed the reconstituted mouse to make a secondary antibody response to an antigen against which the donor cells had been primed. Spleen cells, in other words, contain the memory cells required to generate secondary immune responses. Mitchison primed some mice with DNP-BSA, and others with OVA, and examined, in irradiated mice, the requirements to obtain an optimal secondary anti-DNP antibody response on challenge with DNP-OVA. He found that irradiated mice had to be given spleen cells from mice primed with DNP-BSA, as well as spleen cells from other mice primed to OVA, to generate an optimal secondary anti-DNP antibody response on challenge with DNP-OVA. These observations demonstrate again that an optimal secondary antibody response to a hapten requires cells primed to the hapten and other cells primed to the carrier.[51]

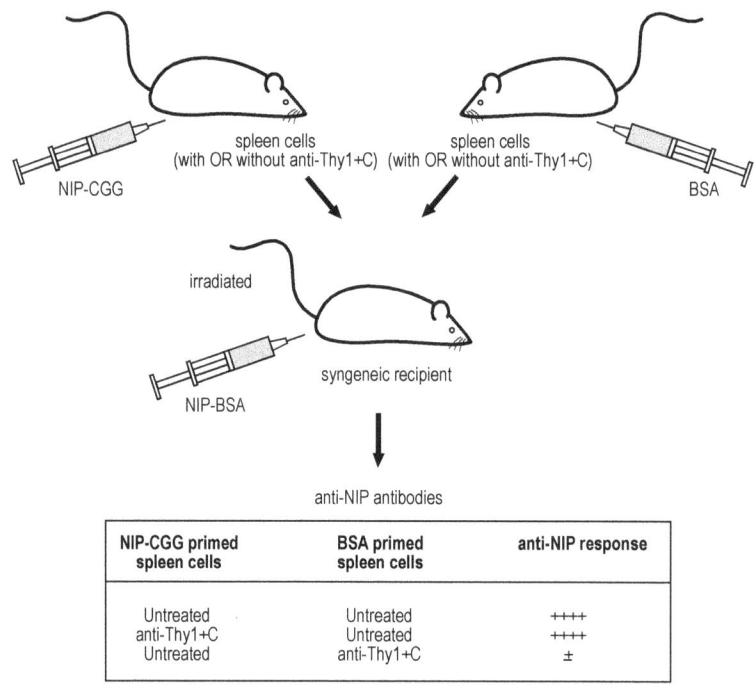

Fig. 14. The carrier-primed cells, required to obtain a secondary anti-hapten response on challenge with a hapten-carrier conjugate, are sensitive to treatment with anti-Thy1 antibody and complement and therefore are thymus-derived cells

Martin Raff, in Mitchison's laboratory, showed that this secondary response to the hapten was abolished if the cells primed to the carrier were treated with anti-Thy1 antibody and complement (C) to remove Thy1+ cells, before being given to the irradiated recipient, see Figure 14. In these experiments, Raff used a different hapten, ***nitro-iodo-phenyl (NIP)***, and a different carrier, ***chicken γ-globulin (CGG)***, but the principles are the same. These experiments show that the cell that recognizes the carrier, and required to obtain a secondary anti-hapten response on a challenge with a hapten carrier conjugate, bears the Thy1 antigen on its surface, i.e. is a T cell.[52]

Given all these experiments, some carried out in the United States, some in Australia, and others in England, the immunological community underwent a collective eureka. It seemed that in both primary and secondary responses, the activation of antibody precursor cells needed antigen-specific T cells. Antibody precursor cells were present in mouse **bone marrow cells** and came to be referred to as **B cells**. What was the role of these T cells in antibody responses?

The nature of the B cell/T helper cell interaction

Neonatal thymectomy of mice renders them unable to make an antibody response to most antigens as adults. However, they are still able to respond to some. These antigens are referred to as *thymus-independent antigens.* Some of these natural antigens, often bacterial polysaccharides, have the feature in common of repeating motifs in their structure, and are therefore referred to as *polymeric antigens.* Other synthetic thymus-independent antigens are also polymeric. It was recognized that this polymeric feature allowed single molecules of these antigens to bind multivalently to the antigen receptors on B cells, thereby aggregating and binding very tightly to them. It seemed to many that the function of T cells might be to bind non-polymeric antigens in a manner to facilitate the interaction of antigen with B cells, in a way that mimicked the interaction of polymeric antigens with B cells. In this case, T cells would not be strictly required to generate antibody responses, as exemplified by the antibody response to polymeric antigens. The T helper cells were envisaged to passively facilitate the presentation of the antigen to the B cell. Their function would be as helpers, and the name T helper cell soon took.

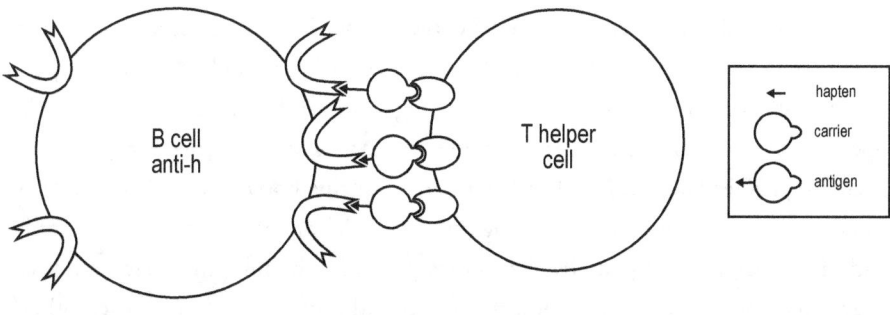

Fig. 15. The Antigen-Bridge Model for the induction of B cells

This view of T helper cell function was widely prevalent and incorporated into a popular model in which the B cell and T helper cell were held together during the activation of the B cell through their mutual binding to antigen. This model is conveniently referred to as the ***Antigen Bridge Model of the B cell/T helper cell interaction*** for B cell activation, see Figure 15.

It was exciting to see diverse experiments coming together into one conceptual framework. There were, however, some aspects of this picture that bothered Mel and me. We had proposed that cell collaboration did not just facilitate the activation of B cells, but that whether collaboration takes place determines whether antigen activates or inactivates the B cell. Most people considering the mechanism of B cell activation did not concern themselves with what circumstances might lead antigen to inactivate B cells. It seemed to us that there must be a more significant role for T helper cells than focusing the presentation of antigen in the form of a repetitive array. Moreover, we thought that antigen could interact *well enough* with the B cell, in the absence of T helper cells, to inactivate the B cell. We considered that the require-ment for a T helper cell could only be mandatory for the activation of the B cell if the B cell received signals during the process of its activation that were both *mandatory* as well as *dependent* on the presence of T helper cells. What could these signals be? We thought two possibilities likely or plausible. The T helper cell could release short range molecules that bound to receptors on the B cell's surface, thereby initiating chemical events inside the B cell in analogy with how neural transmitters function, and/or there could be a requirement for a T cell/B cell membrane/membrane interaction, similarly leading to signaling events inside the B cell. We shall later see that observations subsequently made showed that the activation of B cells involves it receiving both types of communication from the T helper cell.

The Two Signal Model of lymphocyte activation

Jacques Monod was a Non-Resident Fellow of the Salk Institute, as was Francis Crick. They visited the institute for a few weeks every year. Mel had done his post-doctoral studies with Jacques in Paris, making significant contributions to the "lac operon" story, the first elucidation of how the expression of genes is regulated in bacteria. Jacques had received the Nobel Prize for this body of work. Naturally, he came to enquire about what was going on in Mel's laboratory. Not only that, but he had been intrigued for years by the riddles posed by the immune system and had speculated and written on the subject. Jacques and I found we enjoyed each other's company. We had a mutual empathy upon our first meeting. Jacques took a great interest in our model, and was very enthusiastic. He certainly went further than I felt able to. *It is too beautiful not to be correct*, he assured me. Our conversations ranged over many subjects. He appreciated the fact that I, then a mere twenty-six-year-old, would sometimes disagree with him. In fact, he seemed to take a delight in it. Our most intense subject of discussion, and of considerable disagreement, was on the philosophy of science. I always looked forward to his visits.

Jacques made a significant contribution during this first meeting, even though it did not change the way we thought. In hearing about the different possibilities of how we thought the activation of B cells could be made dependent upon the presence of T helper cells, that is through the production of soluble mediators and/or a membrane/membrane interaction, he urged us to devise a readily graspable notation to express the essentials of our view, unclouded by different possibilities. We had used the term "inductive signal" and "paralytic signal" in our first paper in *Nature*. We saw the virtue of Jacques' advice. We therefore began saying in our discussions that the interaction of the antibody receptor with antigen resulted in the generation of *signal 1*, and that the recognition of a second site on the antigen, by the T helper cell, led to the generation of *signal 2*, see Figure 16. We of course envisaged that the generation of signal 1 alone led to the inactivation of the cell. We came to refer to this formulation as the ***Two Signal Model*** of lymphocyte activation. I felt particularly happy about a feature of this scheme. I had thought early on that it would be a nice safeguard if the signal for the activation of lymphocytes included the signal for inactivation. In this case, a cell bearing a receptor unable to mediate the signal for inactivation could not be activated.

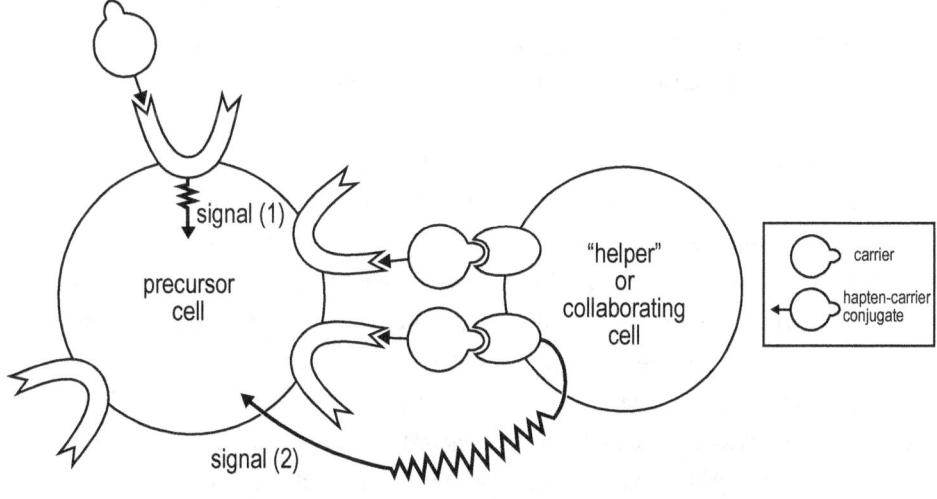

Fig. 16. The original Two Signal Model of lymphocyte activation

We were invited to write an article for *Science* in 1970. Entitled 'A Theory of Self-Nonself Discrimination', it is my most quoted article. We carefully outlined the grounds for the Two Signal Model for lymphocyte activation.[47] It presented our ideas in the most polished form we could devise. We had time to consider and reconsider every word. I know we went through well over twenty drafts, each retyped by Gisela, Mel's secretary. I felt guilty towards her. We explicitly proposed in this paper that the activation of precursor helper T cells also required T helper cell cooperation, and that the interaction of antigen with a single T helper cell led to its inactivation. I stress this because, though our most quoted paper, it is often incorrectly stated that we did not consider what the requirements might be to activate T cells. This proposal regarding the activation and inactivation of T cells is a more detailed and explicit form of the one cell/multiple cell model for the inactivation/ activation of lymphocytes. These proposals were critical to my further theoretical and experimental work. Our proposals accounted for self-nonself discrimination at the level of the mature T helper cell, in a manner consistent with the Historical Postulate. I believe these considerations, although made more than 45 years ago, are central to current major issues in immunology, as I shall later discuss. I feel I should add that this outline, over the last twenty pages or so, of how the Two Signal Model arose, is my perspective. Cohn has given a considerably different account.[53]

Monod as umpire

There were many distinguished visitors at the Salk Institute. The Non-Resident Fellows usually came at the same time every year for about six weeks, and partook in institute seminars and other activities. At the time of these extended visits, other well-known scientists took the opportunity to visit. I believe it was in the summer of 1969 that Baruj Benacerraf, a well-known immunologist, visited us when Jacques Monod was in residence. I knew of Benacerraf and of his work. Indeed, the paper I had come across whilst a graduate student, employing guinea pigs unable to produce antibody to poly-L-lysine (PLL) upon immunization with this antigen, was from Benacerraf's lab. This paper had provided support for my single cell/multiple cell model for the inactivation/activation of lymphocytes. Mel had told me that Jacques and Benacerraf knew each other quite well, as they were related through marriage.

A meeting was arranged with Jacques, Baruj Banacerraf, Mel, and me to discuss immunology. Benacerraf started by declaring that the single cell/multiple cell model could not be correct, as he had evidence against it. Jacques seized the moment. He firmly suggested that Benacerraf outline the design of the pertinent experiment, giving no clues as to what findings were made. Once we had the outline of the experiment clear, Jacques invited me to predict what observations were expected, together with my reasoning. I was happy to comply. When I had made the predictions, Jacques asked Benacerraf whether my predictions were correct. Benacerraf said my predictions were indeed what was found, but the reasoning I had given had to be wrong, in view of another experiment he and his colleagues had done. He said this with some irritation. Jacques requested that Benacerraf outline the design of this further experiment, and invited me to predict what observations were made. We repeated this process three or four times, always with the same result. I then remembered I had to collect a sample from a centrifuge, so I made my apologies and said I would be back in five minutes. When I came back, I found Benacerraf was quite different, and pleasant. We talked about other things than those related to the model. Mel later told me that Jacques had given Benacerraf a dressing down as soon as I had left the room, telling him he should be objective and try to understand our model.

Both Monod and Crick were usually straightforward in discussing science. They indicated when they thought something implausible, though often subtly and with sensitivity, and were, I think, less kind when they thought someone was playing a game. The occasion with Monod and Benacerraf was one of several in which well-known people displayed their difficulties in objectively discussing our model. Another example occurred when Mitchison visited us at the Salk Institute. He gave a seminar on the interaction of B cells and T helper cells. He started by outlining the content of his seminar, adding "but I will not discuss mechanisms, as you have your own high priests here." I had and have a very high regard for Mitchison's contributions to immunology, and am sorry he felt it appropriate to take such a stance.

Analysis of the immune system at the molecular and at the cellular level, and at the level of the system

It is useful to recognize that the nature of the immune system can be analyzed and described at different levels. Molecular explanations or descriptions have to be consistent with the cellular description that is explicitly stated or implicitly assumed. The Two Signal Model of lymphocyte activation is primarily a description at the cellular level. The proposal that signal 2 could be mediated by short range molecules and/or membrane/membrane interactions sets the stage, if correct, for advancing descriptions at the molecular level. In addition, our proposal at the cellular level was consistent with ideas and evidence at a higher level, in that it was consistent with the Historical Postulate. We need a way of referring to observations, hypotheses and descriptions made at this higher level of organization. I shall refer to such ideas or evidence as providing a description at the level of the system. It is often the case that descriptions are first formulated at the highest level and then, as the field advances, they become transformed into lower and more detailed levels. An appreciation of this layered structure is pertinent not only to immunology, but to many disciplines such as physiology and neurobiology. An excessive focus at the molecular level, without a careful consideration of the consequences of these tentative descriptions at higher levels can, I believe, lead to implausible and invalid frameworks.

Concerns arising from the Two Signal Model

The generation of auto-antibodies is easily, and perhaps too easily, explained

There was an aspect of the Two Signal Model of lymphocyte activation that both delighted and troubled me considerably. We have seen that rabbits, made unresponsive to the protein BSA by giving them large amounts of it shortly after birth, could be induced to produce anti-BSA antibody by challenging them with the crossreacting antigen HSA at three months of age, at a time when they were unresponsive to BSA, see Figure 12. We refer to such situations by saying that the *unresponsive state to the "self antigen"*, BSA in this case, can be broken by immunizing with an antigen that crossreacts with this self antigen. How the Two Signal Model accounts for this phenomenon is outlined in Figure 17. Does this type of experiment provide a model for how autoimmunity might actually arise? Observations suggest it does—a conclusion that supports the Two Signal Model.

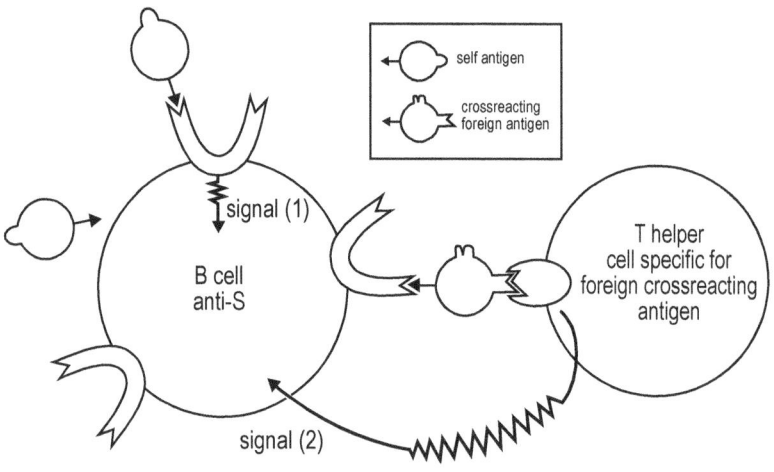

Fig. 17. An explanation, in terms of the Two Signal Model, for the production of antibody against a self antigen S, associated with an immune response to a foreign antigen that crossreacts with S

In one autoimmune disease, Hashimoto's disease, named after its discoverer, the immune system attacks the thyroid gland. The gland produces the hormone ***thyroxine*** that is vital to metabolism and growth. When this disease occurs early in life and is untreated, it results in abnormal development. Luckily, it can be treated if diagnosed early enough by the administration to the patient of synthetic thyroxine. The relatively small molecule thyroxine is produced by the thyroid gland from the considerably larger molecule thyroglobulin. People with Hashimoto's disease often have in their serum antibodies that bind to their own thyroglobulin. It was natural to explore how such *autoantibodies* might be produced. Such an exploration by Weigle was carried out in the 1960s and 1970s with rabbits.[54]

Rabbits, immunized with rabbit thyroglobulin, do not produce antibodies upon this challenge. However, rabbits immunized with turkey thyroglobulin do produce antibodies that react with rabbit thyroglobulin. Turkey thyroglobulin is similar to rabbit thyroglobulin in some parts of the molecule and different in others. The production of antibodies able to bind to both rabbit and turkey thyroglobulin must mean, according to Clonal Selection Theory, that there are B cells in a normal rabbit with antibody receptors able to bind similar *epitopes* present on rabbit and turkey thyroglobulin. These precursor cells cannot be activated by rabbit thyroglobulin, presumably because there are insufficient T helper cells specific for rabbit thyroglobulin, due to partial tolerance to this self antigen. However, these B cells can be activated by turkey thyroglobulin because there are T helper cells specific for those parts of turkey thyroglobulin that are recognized as foreign in the rabbit. These experiments with rabbit and turkey thyroglobulin are in all respects parallel to those described above with BSA and HSA, except that in the latter case, immunologists tried to adapt the rabbit's immune system to regard the foreign antigen, BSA, as self by giving large amounts of it to the rabbits neonatally. In the experiments with thyroglobulin, the rabbit thyroglobulin is a bona fide self antigen.

The relevance of this scheme for how autoimmunity might arise is supported by clinical observation. For example, infection with the bacteria known as *group A streptococci* can result in *rheumatic heart disease*. In this disease, there is an immune attack against heart tissue, associated with antibodies that react with this tissue. These antibodies also react with the streptococci, showing that the streptococci and heart tissue crossreact.[55] These and other observations, in clinical and in experimental systems, were consonant with our proposals as to how lymphocytes are activated and inactivated. This was exciting. However, there was something troubling about all this success.

The paradox

Infection of cells by viruses is common and inevitably leads to an infected cell that is antigenically similar to an uninfected self-cell, except that the infected cell expresses foreign, viral antigens. The virally infected cells crossreact with their uninfected counterparts. The occurrence of sustained responses to such foreign cells, which crossreact with self-cells, would often lead to autoreactivity, according to our framework. This conclusion troubled me. I summarized the paradox of our journey by saying we had developed models for the antigen-mediated activation and inactivation of lymphocytes that accounted for self-nonself discrimination, and our model, as one might hope, accounted for at least some of the conditions under which autoreactive antibodies can be produced. The real trouble was that these conditions would seem to occur frequently, so our explanation of self-nonself discrimination was not that good! Yet our explanation was also supported by its ability to account for the conditions known to result in autoreactivity. This series of thoughts led me to tentatively accept that our explanation of self-nonself discrimination might be correct and to focus on a related question: given the fact that the generation of autoantibodies occurs frequently, how might their detrimental effects be minimized? This underlying unease on my part turned out to yield rewards, to my mind, as it became the spur to further thoughts. As someone once said, just as a field must be plowed before being sown, so must a mind be troubled before being open to a new idea.

The discovery of suppressor T cells: a role in self-tolerance?

Peter McCullagh, working in Australia, gave rats within a day of birth a large dose of SRBC. He was attempting to adapt the rat's immune system to regard SRBC as a self antigen. A high proportion of these rats, when challenged a few weeks later with SRBC, did not produce antibody, though antibody was produced by unexposed rats upon a similar challenge. We refer to these exposed rats as unresponsive. It was natural to initially suppose that these unresponsive rats were in a similar immunological state with respect to SRBC as they were to their own self antigens, as Peter McCullagh and others imagined. Peter McCullagh found that these rats had cells in their lymphoid organs that could inhibit anti-SRBC antibody responses. He demonstrated this in the manner depicted in Figure 18.

Rats were lethally irradiated and all were given a number of spleen cells from naïve, syngeneic rats, which allowed them to produce a copious antibody response to SRBC when challenged with this antigen, see group 3. The generation of this antibody response was inhibited in rats similarly irradiated, reconstituted, and challenged if they also received spleen cells from the unresponsive rats, see group 4. It therefore seemed that these spleen cells from the unresponsive rats inhibited the generation of the antibody response that is otherwise produced by the spleen cells from normal rats. The spleen cells from SRBC-unresponsive rats inhibited the antibody response to SRBC but not to other, non-crossreacting antigens, so the inhibition seemed to be specific. Moreover, if the spleen cells from the unresponsive rats were harvested and the T cells removed from this population, before being given to the irradiated and reconstituted rats, the generation of the anti-SRBC response was no longer inhibited, see group 5. It appeared that the rats are unresponsive because they harbor T cells that inhibit the antibody response to SRBC.[56]

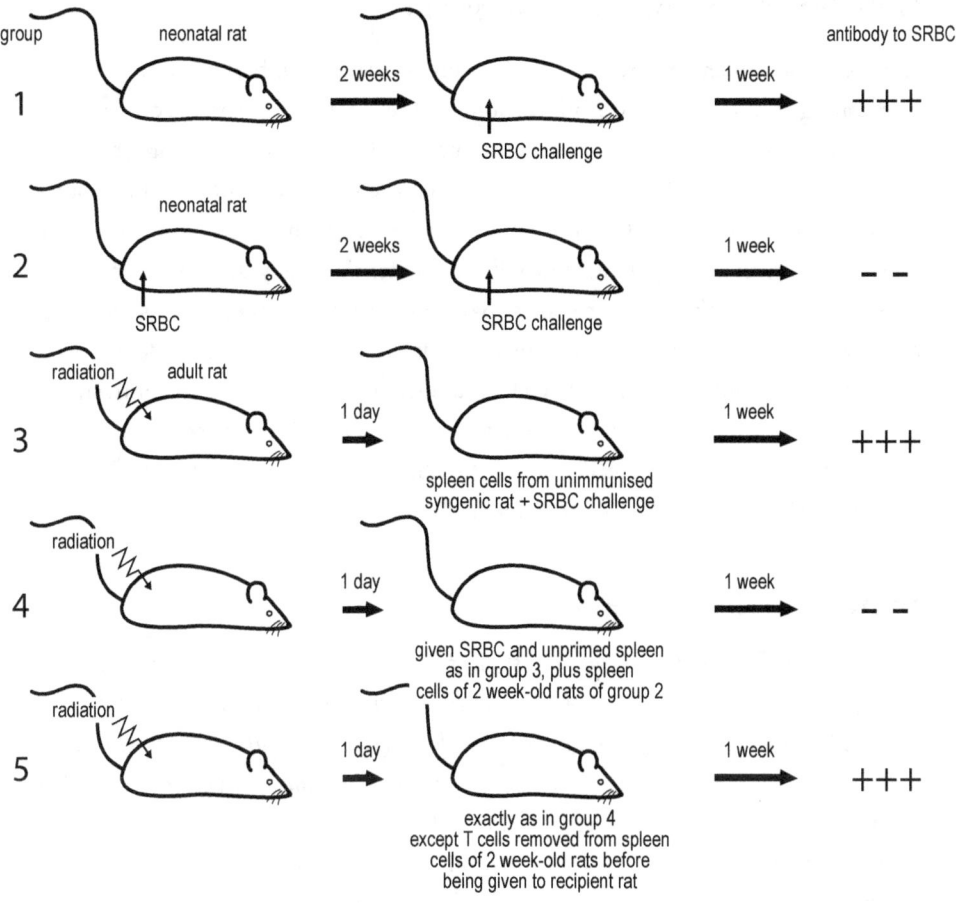

Fig. 18. The demonstration of suppressor T cells in rats rendered unresponsive to SRBC

Observations along similar lines were reported around the same time, in the early 1970s, from the USA laboratory of Dick Gershon.[57] It became accepted that there most probably was a role for *suppressor T cells* in maintaining tolerance to self antigens. How these observations could be squared with the Two Signal Model of lymphocyte activation, and the explanation it provided for self-nonself discrimination, was unclear. Perhaps these experiments were beginning to open up a completely new vista as to how self-nonself discrimination is achieved. This new perspective might undermine that provided by the Two Signal Model. Yet there was also considerable evidence and attractive considerations supporting our model. I explain in the next chapter how these awkward facts spurred me to consider the possibility that these suppressor T cells might have a different physiological role than that of ensuring self-tolerance.

Defining different T cell subsets

Another means of characterizing cells, besides assessing their biological function, is by the antigens they uniquely produce. Some of the most useful markers distinguishing different cell types are antigens on the surface of one cell type that are sparse or absent on another. We have already appreciated the existence and use of two such markers, the surface Ig that acts as an antigen receptor on B cells, and the Thy-1 antigen on thymus-dependent mouse lymphocytes. Virtually all Thy-1-bearing T cells, found in lymph nodes or spleen, have one of two surface markers, denoted as **CD4** and **CD8.** It is found that T helper cells are CD4+, and can be depleted by treating a cell population with anti-CD4 antibody and complement. In contrast, the CTL, generated upon viral infection and able to lyse virally-infected cells, are CD8+. Immunologists found that the T cells able to inhibit antibody responses are CD8+ T cells, and so distinct from CD4+ T helper cells.

Synopsis of Chapter 3

I became intrigued, whilst a graduate student, by the problem of immunological self-nonself discrimination. Immunologists at the time envisaged that an antibody precursor cell could interact in two ways with antigen, one leading to its activation, the other leading to its inactivation. This scenario led to a major question: in what way was the interaction of antigen with the precursor cell different that led to these different outcomes?

I envisaged that a single precursor cell on interacting with antigen is inactivated, whereas its activation requires the antigen-mediated interaction of the precursor cell with other antigen-specific cells. These ideas were appealing for four reasons. First, they were the only explicit model proposed for how these two processes are differentially triggered. Second, they explained why antigens had to be macromolecular in order to be immunogenic, as small antigens would be unable to mediate cellular cooperation. Third, these proposals provided a plausible explanation for how self-nonself discrimination could be achieved. Precursor cells specific for a self antigen would be inactivated as they are generated, one or a few at a time, by virtue of the continuous presence of the self antigen. Precursor cells with receptors unable to interact with self antigens, but able to interact with a particular foreign antigen, would accumulate in its absence; when this foreign antigen later impinged upon the immune system, it could mediate the interactions between the accumulated specific lymphocytes required for their activation. Lastly, they naturally explained a series of observations, dating back to 1961, on the requirements to generate antibody responses in normal animals and animals made unresponsive to an antigen.

I arrived to do postdoctoral studies with Mel Cohn in the latter part of 1968. We decided, within a few months of my arrival, to submit a paper to Nature *on our ideas as then formulated, which was accepted.*

The model we outlined went further than the one cell/multiple cell model for the inactivation/activation of precursor cells. We proposed that the antigen-specific cell, able to help in the activation of the antibody precursor cell, was different from the antibody precursor cell itself in two respects: this cell was envisaged to produce a special class of antibody, we called carrier antibody. We also surmised that the carrier antibody precursor cell, in order to be optimally effective in allowing antigen to activate antibody precursor cells, had itself to be activated.

Our paper was published in November 1968. A number of important publications came out around that time and a bit later, collectively leading to the conclusion that the induction of most primary and secondary antibody responses requires the antigen mediated interaction between the antibody precursor cell, found in the mouse among bone marrow cells and called a B cell, and a helper cell, found in the thymus and called a T cell. Both cell types are present in the spleen. The B cell is the precursor of the antibody producing. What then is the role of the T cell?

Neonatal thymectomy of mice results in the adult mouse having a deficiency of T cells. Such adult mice are unable to generate antibody responses to most antigens. They can, however, generate antibody responses to some carbohydrate antigens that have repetitive, polymeric structures. One molecule of these polymeric, thymus-independent antigens can bind to many receptors of a B cell. These observations led most immunologists to conclude that the T cell is not strictly required to activate the B cell, but just helped non-polymeric antigens to make an array of repetitive determinants with which the B cell could interact. This type of model, shown in Figure 15, is called the Antigen Bridge Model of the B cell/ T helper cell interaction.

Mel and I found this view difficult to accept. We envisaged that the presence and absence of activated T helper cells determine whether antigen respectively activates or inactivates the corresponding B cells. The mandatory role we predicted for T helper cells, in the activation of B cells, could only occur if B cell activation required signals uniquely provided by T helper cells. We suggested that T helper cells generate signals by producing short-range molecules, recognized by receptors on the B cell's surface, and/or by forming a membrane/membrane interaction between the two cells. Later studies showed that T helper cells use both means to facilitate B cell activation.

We proposed in our 1970 Science paper that the signal generated when the antigen-specific receptor interacts with antigen be called signal 1. The generation of signal 1 alone was postulated to result in lymphocyte inactivation. We predicted that activation also required the recognition of a second site on the antigen, leading to the generation of signal 2, mediated by short range molecules and/or by membrane/membrane interactions, see Figure 16. We proposed similar rules govern the activation of B cells and precursor T helper cells.

Many observations are accomadated by this Two Signal Model, in particular the conditions under which autoantibodies are induced. These diverse experiments can be illustrated by one example. Some people suffer from an autoimmune disease, Hashimoto's disease, in which antibodies are produced against the thyroid gland, important in controlling growth and development. Antibodies to thyroglobulin, a prominent protein of this gland, are usually present in individuals suffering this disease. Weigle determined the conditions under which such autoantibodies can be generated. Immunization of rabbits with rabbit thyroglobulin does not result in such antibodies, but immunization with the cross-reacting antigen, chicken thyroglobulin, does. There must therefore be B cells specific for determinants common to rabbit and to chicken thyroglobulin. These B cells are induced by chicken but not rabbit thyroglobulin, presumably because there are sufficient T helper cells specific for chicken but not for rabbit thyroglobulin. The lack of sufficient T helper cells specific for rabbit thyroglobulin presumably reflects a state of partial tolerance. This example illustrates how antigens that cross-react with self antigens have the potential for inducing autoimmunity when only a few anti-self precursor cells remain, or when such precursor cells are generated during the course of the immune response to the foreign, cross-reacting antigen. A real-life situation, where an immune response to a cross-reacting antigen leads to autoimmunity, is rheumatic heart disease. This is precipitated by infection with group A streptococci. These bacteria crossreact with heart tissue!

Progress is sometimes fostered when one is aware of problematical aspects of an appealing framework. Despite the success of the Two Signal Model, there were two troubling aspects as far as I was concerned. The most straightforward trouble came in the form of several different reports in the early 1970s. These studies showed that animals, made unresponsive for the production of antibody by administering antigen to animals under "non-immunogenic" conditions, harbor antigen-specific T cells that inhibit the corresponding antibody response. These observations were interpreted by virtually all immunologists as implying that the activation of anti-self lymphocytes is normally held in check by such inhibitory T cells. It seemed, should this interpretation be correct, that the explanation of self tolerance provided by the Two Signal Model must be at least incomplete, but more probably invalid.

The second trouble was of a more subtle and conceptual nature. We had tried to explain how self-tolerance is achieved. Our Two Signal Model provided such an explanation and accounted for how autoantibodies can be induced by the impact of a foreign antigen that cross-reacts with a self antigen, as just outlined. However, these successes had their troubling aspects. Virally-infected self cells, and cells infected by other intracellular pathogens, cross-react with uninfected self cells, and therefore have the potential to induce autoimmunity. Such infections are frequent and so autoimmunity should be common. Indeed, there are reports that, with sensitive techniques, autoantibodies can often be detected specific for diverse self antigens. Was such autoreactivity damaging and, if so, how was this autoimmunity tolerated?

Chapter 4

A THEORY OF IMMUNE CLASS REGULATION

Setting the scene

At the same time as some immunologists were focusing on the nature of self-tolerance, others were making significant observations on immune class regulation. The first observations demonstrating the existence of distinct classes of immunity were made at the level of the whole animal. We have already seen that some immunity is mediated by antibodies, and that other kinds or classes of immunity are mediated by cells. Robert Koch had found that individuals with tuberculosis expressed **delayed-type hypersensitivity (DTH)** to a protein extract, **purified protein derivative (PPD)**, prepared from the causative organism, *Mycobacterium tuberculosis*. Landsteiner and Chase later showed, as we have seen, that this state of DTH can be transferred from a sensitized to a naïve animal by the transfer of immune cells but not by the transfer of immune serum. Such observations led to a recognition that **DTH** is an example of **cell-mediated immunity (CMI)**. We have also seen that viral infection also usually leads to the generation of **cytotoxic T lymphocytes (CTL)** that can lyse virally-infected target cells. These CTL constitute another limb of cell-mediated immunity.

It had also become apparent in the 1950s and 1960s that different circumstances of immunization are required to generate different classes of immunity. This generalization is illustrated by Salvin's systematic study of the 1950s.[19] Salvin mapped the kinetics of expression of CMI in the form of DTH, and of antibody production following a single challenge with a protein antigen, as we have noted, see Figure 8 of Chapter 1. To recall, the tempo of the responses observed is affected by the dose of the administered antigen. A high dose results in a rapid antibody response, sometimes with barely detectable or undetectable expression of DTH. As the dose of antigen is lowered, the production of antibody is less rapid and is preceded by a phase where DTH is uniquely expressed. The expression of DTH decays as antibody is produced. An even lower dose of antigen results in an exclusive DTH response.

Observations show that CTL, also an expression of CMI, are usually generated before substantial antibody can be detected. It is particularly interesting, for example, that individuals infected by HIV-1 usually produce a CTL response against HIV-1 infected cells before anti-HIV antibody can be detected in their blood.[58] Once such antibody is detectable, the individual is said to have **seroconverted**. Individuals infected by HIV-1 are relatively symptom-free before and for a limited time after they seroconvert. This period, which can vary considerably in length between different individuals, is referred to as the "honeymoon period" due to its relatively symptom-free state. This is a fascinating finding to which we shall return later. It seems to some, including me, to provide a clue as to how one might achieve effective vaccination against and treatment of AIDS in the early stages of the disease.[6]

Variables affecting the class of immunity generated

We have just reviewed how the class of immunity expressed at one time, following primary immunization, depends upon the dose of antigen administered and the time after immunization at which immunity is assessed. In addition, there are other observations on how different variables of immunization affect the class of immunity induced. I became interested in the early 1970s in understanding how antigen interacts with cells of the immune system to determine the class of immunity induced, and in thinking about what physiological purpose the generation of such distinct classes of immunity, and their differential regulation, might serve. A major impetus for my speculations was a knowledge of the variables of immunization that affect the class of immunity induced. I thought, and still think, that any viable proposal, as to how antigen interacts with cells of the immune system to determine the class of immunity generated, must account for these variables. I believe that the framework I developed in the early 1970s accounted for the variables then known,[59] as well as some that have since become apparent.[60] Other contemporary proposals do not seem to me to have this property. These are important issues. They bear on the application of basic science to realize the prevention and relief of much human disease and suffering. It is for these reasons that I wish to outline what these variables of immunization are. This will provide the context for discussing ideas on the nature of the **decision criterion** that determines the cell-mediated/humoral nature of the immune response.

Past exposure to the antigen

We have seen how different conditions of immunization, such as those employing different doses of antigen, can lead to the generation of distinct classes of immunity. It seems a rather general phenomenon that, as antibody is produced, the expression of DTH decays. A number of investigators, starting in the mid-1960s, examined how pre-exposure of an animal to antigen, in a manner that results in substantial antibody production, affects the animal's ability to subsequently generate a DTH response. The type of experiment performed is outlined in Figure 19. Animals are first given an antigen challenge that results in an antibody response, for example a challenge with a substantial dose of antigen. They are then rested for one week, say, after which they are given a challenge that induces DTH in normal animals, for instance a low dose of the antigen. It is found that DTH can no longer be induced in the animals pre-immunized to result in the production of antibody. The immune response appears to be locked into an antibody, or humoral, mode. This phenomenon was termed *immune deviation* by Asherson and Stone in 1965.[61] We refer to this state as a state of **humoral immune deviation** to indicate, in the name, the direction in which immunity is deviated.

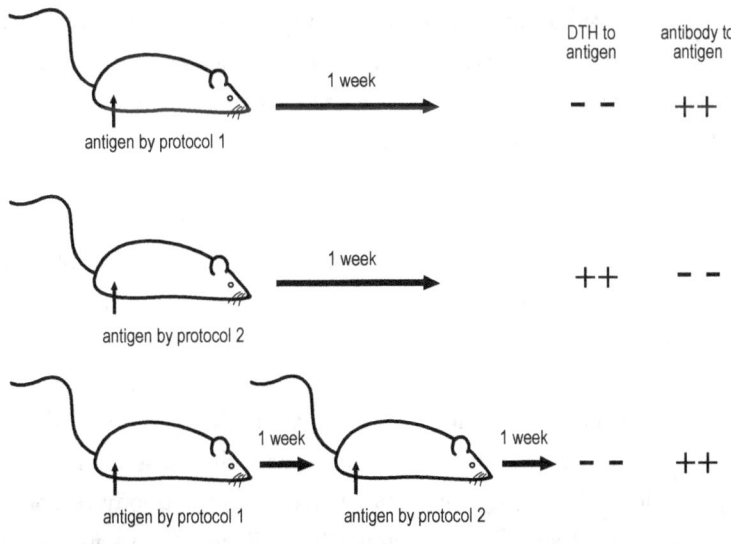

Fig. 19. Outline of experiments revealing the phenomenon of humoral immune deviation

In the late 1960s and early 1970s another form of immune deviation was discovered. Parish and colleagues found that it was possible to lock the immune response into a cell-mediated mode,[62] a state we refer to as **cell-mediated immune deviation**. However, to set the scene for this discovery, it is helpful to digress for a moment to describe some other major findings.

Low-zone and high-zone paralysis

Immunologists first imagined that the immune system would adapt to regard an antigen as self if the antigen is first present early in development and is continuously present thereafter, an idea referred to as the *Historical Postulate*.[63] They then found that unresponsiveness could sometimes be achieved by giving massive amounts of antigen just after birth. This unresponsiveness was demonstrated by subjecting the animal, at a few weeks of age, to a challenge of antigen that induced a brisk antibody response in naïve animals, as depicted in Figure 11 of Chapter 3. We have already come across two experiments of this type: the administration to rabbits, just after birth, of the protein BSA[45] (Figure 12), and the administration to new born rats of SRBC,[56] (Figure 18). Studies in neonatal mice and rats were easier to realize than those in the corresponding fetuses that are only readily accessible in birds, as birds develop in eggs. In time, immunologists began employing not only neonatal animals but also young animals in their attempts to develop what they imagined were experimental models of self tolerance.[40]

A series of influential experiments was carried out by Mitchison in the mid-1960s. He exposed *immunocompetent* mice to repeated doses of the antigen BSA, given a few times a week. After a considerable course of such exposure, in which different mice received different amounts of antigen, varying over about a 10,000-fold range, he rested the animals for a short while, and then examined their antibody response following a challenge of BSA that induced an antibody response in naïve animals. The observations Mitchison made are summarized in Figure 20.[64] Note that there are four distinct situations we can distinguish from the observations depicted. First, there are those mice that did not receive any antigen before the challenge, indicated by a "1". This response is a primary antibody response to BSA. Second, there are those mice that are pre-exposed to a medium dose of the antigen, and that make a higher antibody response to the challenge, indicated by a "3". These mice became primed for a secondary antibody response by repeated pre-exposure to a medium dose, which was "immunogenic" for an antibody response. Lastly, there are those mice pre-exposed to a considerably lower or higher dose than the medium, immunogenic dose of antigen; these mice make *smaller* antibody responses than a primary response, indicated by a "2" and a "4" in Figure 20. An animal rendered unresponsive to an antigen because of pre-exposure to it was said at the time to be paralyzed with respect to its ability to produce antibody to the pertinent antigen. Thus the diminishment in the primary antibody response to BSA, consequent to the repeated exposure to very low and very high doses of antigen, came to be called *low-zone* and *high-zone paralysis*.

These observations, made at the level of the system, were and are of outstanding importance for three reasons. First, in time, the possible generality of these observations became more plausible. Second, their explanation requires quantitative considerations, and so any explanation of the basis of low-zone and high-zone paralysis must introduce quantitation to the field. Third, these observations provide critical clues as to how the immune response is regulated, if only one can correctly appreciate their physiological significance. This point is pivotal. The most prevalent interpretation, espoused by many for decades and still entertained by most of the rare individuals still cognizant of these observations, is the one first put forward by Mitchison. I feel it is wrong. Mitchison suggested that low-zone and high-zone paralysis corresponded to states of self tolerance.[40] I felt and feel this proposal to be misleading, for reasons to be shortly discussed.

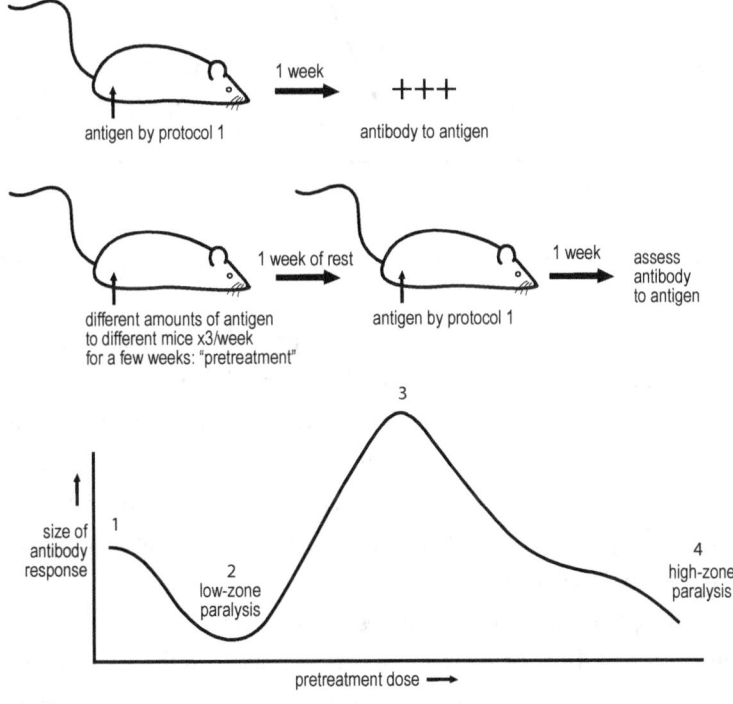

Fig. 20. Outline of experiments revealing low-zone and high-zone paralysis for the induction of antibody.

Low-zone and high-zone cell-mediated immune deviation

Chris Parish subsequently carried out similar studies to Mitchison's, except he employed rats instead of mice, and a major fragment of a protein called flagellin as the antigen rather than BSA. Chris also included an examination of an additional and critical parameter in his studies. He looked not only at the level of antibodies produced following the final challenge, but he assessed upon the challenge the magnitude of the state of flagellin-specific DTH. His findings are illustrated in Figure 21.[62] What this study demonstrated is that low-zone and high-zone paralysis for antibody formation are associated with a state of cell-mediated immunity in the form of DTH. Thus, a state of DTH could be associated with unresponsiveness, or at least partial unresponsiveness, for the generation of antibody. The immune system was locked, or partially locked, into a cell-mediated mode. We refer to this as a state of *cell-mediated immune deviation.*

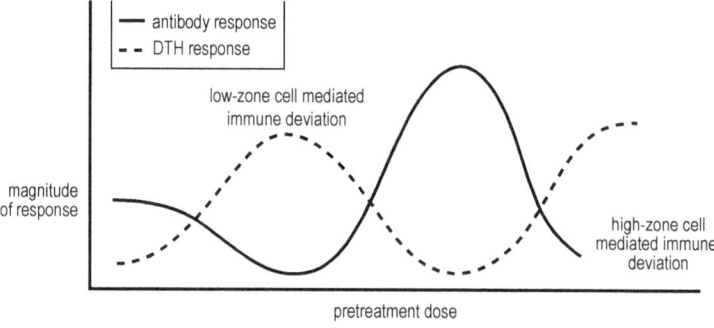

Fig. 21. Parish's experiments demonstrating low-zone and high-zone cell-mediated immune deviation

I found Chris' studies fascinating. They had a great influence on my thinking and led me to go to Canberra, Australia, where Chris worked. I think it a pity their impact on immunologists has not been greater. It was clear to Chris and to me that low-zone and high-zone paralysis were misnomers, and that these phenomena would be more appropriately referred to as *low-zone* and *high-zone cell-mediated immune deviation.* However, for many decades after Chris had published his observations, leading immunologists held the phenomena of low-zone and high-zone paralysis as critical to understanding self-tolerance. It is interesting to summarize the pros and cons for these views. It is tempting for me to make the case that, had the alternative views as to the physiological significance of low-zone and high-zone paralysis been thrashed out amongst immunologists at this time, we might now have an effective vaccine against, for example, AIDS. Naturally, this is a subject to which I shall return.

There were several reasons why I felt that reflection should make anyone skeptical that low-zone and high-zone paralysis were states corresponding to the state of tolerance to self antigens. First, tolerance to self is surely a state of unresponsiveness for all classes of immunity; as originally defined by Mitchison, low-zone and high-zone paralysis were states of unresponsiveness for antibody production, and he did not look at other classes of immunity, as Parish did. When Parish did so, he found that unresponsiveness for the production of antibody was associated with the expression of DTH to the antigen. Second, it was often supposed and stated that, in view of the phenomena of low-zone and high-zone paralysis, the concentration of all self antigens must exist in either a low- or high-zone range. This idea seems implausible in the extreme. There cannot be such constraints upon the concentration of all self antigens in order to solve the problem of self-nonself discrimination. The third point is related to the second. The phenomenon of low-zone and high-zone paralysis refers to a situation where antigen is given to an *immunocompetent animal*. Thus these phenomena, if really relevant to self-tolerance, would correspond to *establishing* unresponsive states. However, all self antigens, or at least the large majority, are present before immune competence arises, so the critical question with respect to self tolerance is not how is self tolerance *established*, but how is it *maintained*? The immune system does not have the ability to respond to most self antigens at any time, so there is no need to establish unresponsiveness, just to maintain it. It seems most plausible that maintenance of tolerance should be ensured by any concentration of a self antigen, so long as it is above a minimal level. This is indeed the case for maintaining tolerance as envisaged in the Two Signal Model of lymphocyte activation.

In addition to these conceptual concerns, Parish's interpretation of his observations could be related to other findings. For example, Salvin had shown in the 1950s that amounts of antigen subimmunogenic for an antibody response induced DTH exclusively,[19] as we have already discussed. Such observations were consistent with the existence of low-zone cell mediated immune deviation.

The nature of the antigen affects the class of immunity induced: the Pearson/Raffel generalization

We have seen that some antigens can induce both antibody and cell-mediated immunity, in the form of DTH, depending upon the circumstances of immunization. The dose of antigen administered is a critical variable in determining the class of immunity expressed at a particular time. Pearson and Raffel made an important generalization in the 1960s. They pointed out that certain antigens, either small in size, or larger but predominantly identical to self antigens and therefore minimally foreign, were only able to induce DTH. Moreover, when these antigens were conjugated to a more foreign carrier able to induce antibody, the conjugate was also able to induce antibody, some of which reacted with the carrier and some with the small, minimally foreign molecule.[65] There must therefore be, according to the Clonal Selection Theory, B cells specific for the small or minimally foreign molecule. The generalization made by Pearson and Raffel is, I think, very important. I shall refer to it as the *Pearson/ Raffel generalization.*

Circumstances leading to a Theory
of Immune Class Regulation

An insight into the physiological advantage to the host of an immunological phenomenon provides a wonderful springboard for speculation. If valid, the insight provides clues as to how the existence of the phenomenon might be regulated. In the
context of immune class regulation, one might well first ask why are there distinct
classes of immunity, and why is there a tendency for their expression to be exclusive?

My interest in how immune class regulation is achieved reached an intensity
sometime in 1972. There was a period of about six weeks of feverish activity. I
was living by myself. My first wife, Elin, being of Norwegian descent, was taking a
prolonged vacation in her home country, and so I was completely free to spend the
hours of every day as I wished. This surely helped. We were living in a small house
in a small village called Solana Beach, north of La Jolla in southern California, and
ten minutes by foot to the Pacific Ocean. During these six weeks, I got up before
dawn every day, took an hour-long walk along the beach, during which I watched
the sun rise, and worked continuously, besides brief breaks for eating, for the next
twenty or so hours. The cycle then repeated itself after four hours of sleep. The ideas
that formed during this time fitted together so beautifully, after much exploration of
different paths, that I became convinced the patterns I discerned had considerable
validity. This was more than I could vouch for the Two Signal Model. This was the
most intense mental activity I have ever experienced. The ideas I formed would be
the basis for much of my work for the following forty years, yet I never managed
to express them in a way that they became known and seriously considered by the
immunological community, though not through want of trying. Personally, I feel
that this theory is as fundamentally significant as, and probably of greater practical
importance than, the Two Signal Model of lymphocyte activation. My failure to
communicate my vision has caused me distress over the decades.

In attempting to explain these ideas, I think it useful to describe the principal
considerations that went into their formulation in a narrative style. I hope this style
still allows an appreciation of the internal consistency or harmony of the proposal.
This harmony convinced me the theory had a certain degree of validity.

Thoughts on the potential physiological
importance of different classes of immunity

Antibody effector function

An important starting point was my coming across a study of the 1960s, aimed
at understanding how antibody must bind to red blood cells in order to facilitate

the "activation" of complement. Such activation results in the punching of holes into the membrane of the red blood cells, causing their lysis. The study revealed that two molecules of IgG antibody had to bind close together on the red cell surface, forming an *IgG doublet*, in order to facilitate the binding of the first component of complement to this doublet, and so initiate the complement cascade. It was found that *several hundred thousand* molecules of IgG antibody had to bind to the surface of a red blood cell for there to be a reasonable chance that an initiating IgG doublet be formed.[66] These requirements seemed, from one perspective, to result in an inefficient mechanism. It also seemed unlikely to me that this inefficiency was mechanistically inevitable; nature could surely have made a more efficient mechanism if such efficiency was advantageous.

Antibody versus cell-mediated effector function

On reflection, my discovery of this cardinal finding triggered three thoughts. Is there an advantage to the bluntness of this weapon of attack? Second, if antibody could not be effective in destroying a cell when recognizing only relatively few foreign sites on the cell's surface, would such a minimally foreign cell induce antibody under the best of circumstances? My guess was probably not. What then would be induced, cell-mediated immunity perhaps? This line of reasoning only made sense if the activated cells of cell-mediated immunity were effective against such a target cell bearing fewer foreign sites.

These musings were given tentative support from different directions. The year was 1972, and there had been a tremendous blossoming in the 1960s of work in tumor immunology in animal models of human cancer. Many of the tumors studied were caused by retroviruses that transform infected cells into a malignant phenotype, or had been induced by carcinogens. It had been found in these systems that cell-mediated immunity was usually required to contain tumors, and that antibody was more predominant in animals suffering from tumor progression.[67] Indeed, it was claimed that anti-tumor antibody could block the killing of tumor cells by cell-mediated effector cells. One of the founders of tumor immunology, George Klein, expressed views then prevalent in an address in 1968. The quotation refers to how effective vaccination against tumors might be achieved: "It will be most important to establish what variables of immunization... dosage, route of administration, and timing are critical to achieve the objective, which seems to be a stimulation of host cell-mediated rejection with minimum risk of antibody-mediated enhancement".[68]

However, there was no discussion as to why cell-mediated immunity but not antibody was effective against these tumors. It seemed likely to me that tumor cells were minimally foreign when compared to bacteria that grow extracellularly, such as diphtheria bacilli, against which antibody was highly effective. It was while these tentative thoughts were forming that I came across the Pearson/Raffel generalization: the administration of minimally foreign antigens could generate only cell-mediated but not humoral responses![65] This report struck me forcefully, and stimulated me to consider seriously the hypothesis that only cell-mediated immunity could be effective against minimally foreign cells and minimally foreign antigens. Antibody-dependent mechanisms would only be effective against more substantially-foreign cells and antigens. Why should this be?

The overall pattern of the regulation of cell-mediated and humoral responses

Now came a series of ideas that I found increasingly attractive as I lived with them. They were related to my unease concerning the frequent occurrence of autoantibodies that I have already described. I thought that a first rule of immune class regulation might well be, on the basis of physiological need, that *the immune system should make, as rapidly as possible, an immune response that can be effective against the foreign invader*. Thus, minimally foreign cells or antigens must generate a cell-mediated response. Moreover, under these circumstances, antibody would not only be ineffective but might interfere with the ability of the activated cells, of cell-mediated immunity, to attack their target cells. It made sense that, when cell-mediated immunity was uniquely needed, the production of antibody was inhibited.

Consider now a cellular antigen with many foreign sites. In principle, both cell-mediated immunity and antibody could be effective and so it initially seemed that either might be generated. Would there be a physiological advantage as to which type of immunity is induced? Suppose the foreign cell crossreacts with some self-cell, such as the case already discussed, in which the foreign cell is a self-cell infected by a virus. According to our Two Signal Model, and supported by observation, this could give rise to autoreactivity.

Now the advantage of the inefficiency of antibody in attacking a cell is clear: if some autoantibody is induced, it would be benign unless the antibodies recognized antigens expressed so densely on a self-cell that the antibody could activate effector functions. Moreover, the generation of cell-mediated autoimmunity, by the cross-reacting antigen, would often be more damaging than the corresponding antibody. Thus, if antibody is able to deal with the foreign invader, there seems to be a disadvantage in generating cell-mediated immunity, as this only increases the damaging consequences of any autoreactivity induced. These considerations provided a teleological reason for why strong antibody responses are associated with suppression of cell-mediated immunity, and why humoral immune deviation exists. These thoughts led me to consider a second rule: *Do not make the weapons of attack more vicious than necessary to satisfy rule #1, as this will increase the debilitating consequences of any auto-reactivity generated.*

Finally, this teleological picture also makes sense in terms of the way the immune response against most antigens evolves with time, going through a cell-mediated phase before a predominant humoral response occurs, as delineated by Salvin (Figure 8 of chapter 1). It takes time for naïve and rare antigen-specific B cells to multiply and differentiate. As many IgG antibodies have to bind a target cell to activate effector functions, low amounts of IgG antibody cannot be effective. Thus, shortly after a foreign cell or invader impinges upon the immune system, before sufficient time has elapsed to produce substantial amounts of IgG antibody, the small amounts of IgG antibody that could be produced would be ineffective. At this early time, according to rule # 1, we must make a sharper response, i.e. a cell-mediated response, that is up to the task of effectively attacking the invader.

In addition, IgM antibody is produced before IgG antibody, and its production is most often down-regulated as IgG production becomes substantial. This also made sense, as IgM is generally more effective than IgG antibody in binding antigen and in activating effector functions. For example, the binding of just one IgM molecule in the correct manner to a red blood cell or a bacterium can trigger the complement cascade. Thus these teleological considerations also explained why it is physiologically advantageous that immune responses to foreign antigens go through a cell-mediated phase before antibody is produced, and why the production of IgM antibody straddles maximal expression of cell-mediated immunity and of IgG antibody production. The switch from a cell-mediated to humoral phase as the immune response evolves minimizes the damaging consequences of any autoreactivity, whilst allowing an effective immune response against the foreign invader.[6, 59, 69]

The basis of cell-mediated and humoral immune deviation

More immunologists in the early 1970s were studying the regulation of antibody than of cell-mediated responses. We have seen how Peter McCullagh, and Richard Gershon and colleagues, had found conditions under which it was possible to render rats, in the case of McCullagh, and mice, in the case of Gershon, unable to produce an antibody response to SRBC. It was demonstrated in both rodent systems that this unresponsiveness was associated with inhibitory, "suppressor T cells". These unresponsive states were, at the time, almost universally interpreted as an experimental state corresponding to that of self-tolerance.

I thought this view might be mistaken for two reasons. First, SRBC are highly immunogenic, and it is not that easy to generate unresponsiveness in newborn rodents to even weak antigens. This led me to consider that these experimental models might reflect a different kind of unresponsive state than that corresponding to self tolerance. I was, of course, loathe to accept the prevalent view, as I could see no easy way of reconciling it with our model of lymphocyte activation/inactivation and its explanation of self tolerance. A most striking finding was in line with my uneasiness. It was found in both systems that the unresponsive rodents are strongly primed to make an exaggerated antibody response to SRBC. It appeared that their anti-SRBC B cells have been partially activated and that a lid has been put on these B cells to restrain their differentiation into antibody producing cells. Both Peter McCullagh and Dick Gershon's group found ways to break the unresponsive state, resulting in copious production of antibody to SRBC at a much greater intensity than occurs in a primary response.[56,70] This breaking of the unresponsive state was achieved in the Gershon laboratory by immunizing the unresponsive mice with **horse red blood cells (HRBC).** HRBC crossreact with SRBC to the tune of about 10%, meaning that about 10% of the anti-HRBC antibody, raised upon immunizing a naïve mouse, would bind to SRBC. Conditions were found under which the immunization of the unresponsive mice with HRBC produced a very rapid and substantial antibody response to HRBC, but almost *all* this antibody could bind to SRBC.[70] The anti-SRBC B cells in the unresponsive mice were primed, poised to produce antibody! It seemed highly unlikely that B cells specific for self antigens were normally primed in this manner.

Consider, however, what happens during the course of a normal immune response to SRBC. This had been studied by George Mackaness, who confirmed for SRBC what Salvin had found for protein antigens: an exclusive state of DTH is usually expressed before antibody is produced, and the administration of sufficiently low doses of SRBC only induces DTH.[71]

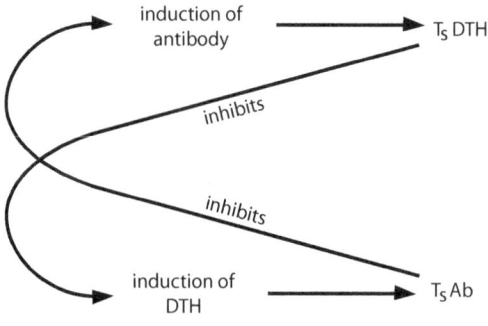

Fig. 22. The proposal for how two types of suppressor T cell might account for the exclusivity of cell-mediated and humoral responses

However, during the course of the exclusive DTH phase, it seemed most likely that the antigen-specific B cells multiplied in preparation for the antibody phase.

I therefore thought that the unresponsive states studied by Peter McCullagh and Richard Gershon might well reflect a state of cell-mediated immune deviation to SRBC; the *suppressor T cells* they identified had the role of inhibiting the primed antibody response. I imagined this inhibitory cell to be coordinately generated with cell-mediated effector cells, and to become more effective, as a suppressor T cell able to suppress the production of antibody responses, if there was a chronic cell-mediated response. Thus, I thought that the low-zone and high-zone states of cell-mediated immune deviation, observed by Parish, were also probably critically associated with the generation of similar inhibitory T cells. I shall for convenience refer to these T cells, that *suppress* antibody responses, as *TsAb cells*.

I similarly proposed that strong antibody responses, associated with humoral immune deviation, involved the generation of T cells that inhibit or suppress the generation of DTH responses. I shall for convenience refer to these T cells as *TsDTH cells*. I was unaware of any indirect evidence for the existence of such cells. However, some antigen-specific mechanism had to account for the decline in DTH as the antibody phase developed. These proposals are expressed in Figure 22.

The "decision criterion" determining whether antigen induces cell-mediated or humoral immunity: The Threshold Hypothesis

The considerations just described bear on events that occur late during the course of a particular immune response. Chris Parish had to immunize rats over weeks to establish low-zone or high-zone cell-mediated immune deviation. Humoral immune deviation only becomes apparent once antibody is produced. The proposal that cell-mediated and humoral immune deviation are respectively associated with TsAb and TsDTH cells does not address the question of how antigen interacts with cells of the immune system to determine, in the first place, whether an antibody or cell-mediated response is generated. We call this decision-making process, embedded in the way an antigen mediates the interaction between cells of the immune system to result in different classes of immunity, as the *decision criterion* determining the class of immunity induced. The decision criterion determines whether antigen induces, under defined circumstances, a cell-mediated or a humoral response. I attempted to explore mentally what might be the basis of the decision criterion.

I was first struck by the Pearson/Raffel generalization that minimally foreign antigens could only induce cell-mediated immunity, and the potential physiological significance of this proposition, in the sense that antibody could most probably not be effective against such antigens. How could the immune system measure the number of foreign sites on an antigen? Everyone at that time believed, and most still do, including myself, that the large majority of lymphocytes specific for self antigens are either killed or rendered non-functional. One clear difference then between an antigen with few foreign sites and one with many was the number of lymphocytes specific for these antigens in an animal. It seemed to me that the different numbers of lymphocytes specific for minimally foreign and more foreign antigens must be a central factor in determining the class of immunity an antigen induces.

The way a consideration develops depends critically upon one's underlying assumptions. I then tentatively believed, and I now more firmly believe as a result of extensive observations, that the optimal activation by antigen of virtually all lymphocytes requires lymphocyte cooperation. In particular, as discussed earlier, I believe the optimal activation of pTh cells requires the antigen-mediated interaction with Th cells or, in modern terms, that the optimal activation of CD4 T cells requires or is facilitated by CD4 T cell cooperation. It was within this context that my ideas formed.

I had thought a great deal about the quantitative aspects of the ideas I was considering. I had made simple models of cellular interactions, and put them into mathematical form.[48] In looking back at these papers now, after several decades, I realize that the mathematics must have constituted an effective prophylactic measure, ensuring few read these papers or took my proposals seriously. I remember Mel relaying to me complaints from colleagues he met at meetings about my mathematics. Despite this, although I regret that I did not manage to make my ideas more accessible, I do not regret devoting serious time to thinking quantitatively. It prepared me to take seriously observations with a quantitative aspect to them. The Pearson/Raffel generalization was a quantitative statement, low-zone and high-zone cell-mediated immune deviation were phenomena only describable in quantitative terms, and the requirements for IgG antibody to signal the activation of complement also had quantitative aspects.

Consider an antigen with few foreign sites and for which there are, in the intact animal, relatively few CD4 T cells. According to the Two Signal Model and the assumptions I tentatively held, the induction of any response must first involve CD4 T cell collaboration. In this case, such cooperation will initially be tenuous, even in the presence of an optimal amount of antigen to mediate this interaction, due to the relative scarcity of the CD4 T cells. Consider now the case of an antigen with many foreign sites, for which there are considerably more CD4 T cells available for cooperation. In the presence of an optimal level of antigen to support CD4 T cell cooperation, such cooperation will be relatively robust. I thus entertained the idea that tenuous CD4 T cell cooperation gave rise to DTH-mediating cells , i.e. to T_{DTH} *cells*, and robust cooperation to T helper cells able to help B cells to produce antibody, i.e. to the generation of T_hAb *cells*. This idea, that different thresholds of CD4

T cell cooperation are required to generate T_{DTH} and T_hAb cells, is referred to as the ***Threshold Hypothesis,*** see Figure 23

This concept became more appealing with the realization that it accounted for other quantitative facts. First, as antigen impinges upon the immune system, antigen-specific CD4 T cells are activated to multiply and so become more numerous; so long as the antigen level is sustained, the strength of CD4 T cell cooperation would increase, and so we can see why the immune response evolves in time from a DTH into an antibody mode. This will inevitably happen, according to the assumptions we have made, unless TsAb cells act to inhibit the further induction of CD4 T cells, and the response consequently becomes locked into a cell-mediated mode. The Threshold Hypothesis accounts in this way for Salvin's findings on how, after antigen impact, the class of immunity evolves with time (Figure 8 of chapter 1).

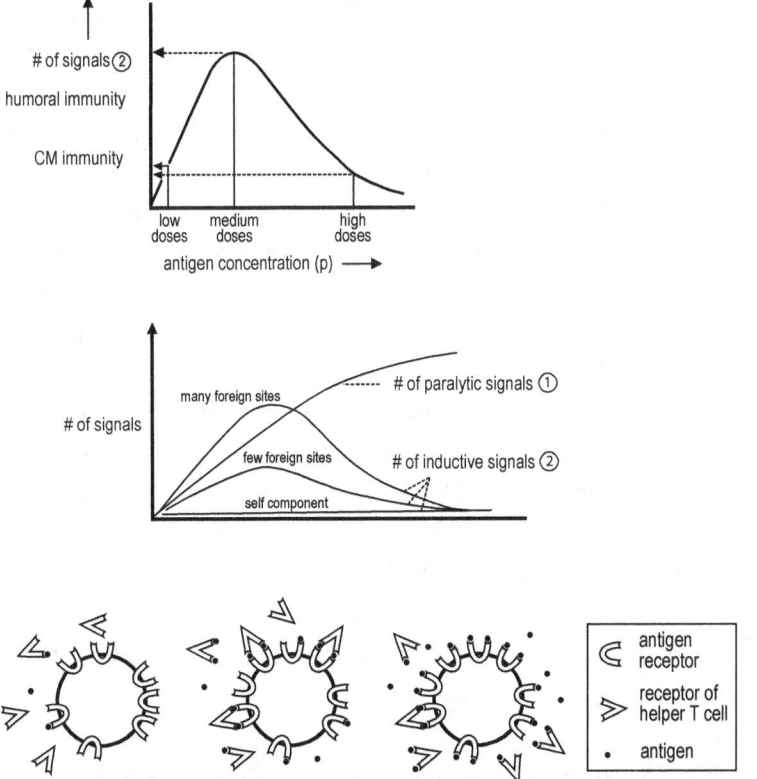

Fig. 23. Two ways of expressing The Threshold Hypothesis, modified from articles published in 1974 and 1977.

Consider what happens when we change the amount of a more foreign antigen from that optimal for CD4 T cell cooperation. As the antigen level is decreased or increased, the level of CD4 T cell cooperation decreases, somewhat similarly as precipitates do not form in the presence of amounts of antigen much lower or higher than the amount of antibody present, see Figure 3 of Chapter 1. The strength of CD4 T cell cooperation will decrease until it fails to reach the threshold required to generate ThAb cells, see Figure 23. In this case, T_{DTH} cells will be generated and, if the CD4 T cell cooperation is sufficiently sustained at this level, the generation of TsAb cells will be sufficient to lock the immune response into a cell-mediated mode. The Threshold Hypothesis thus accounts for how the dose of a foreign antigen affects the class of immune response generated, as delineated by Salvin (Figure 8). The hypothesis also accounts for Parish's observations on low-zone and high-zone cell-mediated immune deviation (Figures 20 and 21).

I also proposed that the optimal induction of IgM antibody was favored by the formation of an intermediate level of CD4 T cell cooperation, see Figure 23. This proposal explains why the production of IgM antibody usually straddles the generation of the predominant DTH and IgG phases of the immune response. The Threshold Hypothesis thus appeared, and still appears to me, to account for all of the variables of immunization known to affect the class of immunity induced, and is consistent with the overall pattern observed as to how the generation of different classes of immunity is generated. This pattern was envisaged to provide a means of fighting forein invaders effectively whilst minimizing the damaging effects of autoreactivity. I felt then, and still feel some forty years later, that these patterns are unlikely to be fortuitous. Moreover, new and striking findings were made consistent with this theory. We developed strategies of vaccination and treatment based upon the theory that show remarkable efficacy. I had had an intense period of about six weeks during which the theory was formulated. I came to the end of my intense activity through the writing of a first draft of a paper outlining the ideas, and giving a rather long talk of a few hours to the members of Mel's lab. One of these was a graduate student, Bob Coffman, who later became a well-known immunologist.

I was exhausted. I had a great need to use my hands, so I took pottery lessons from a good potter as a way of decompressing. I did not make very good pots; however, I had the most sublime dreams in which I was a complete master of the art of throwing the most beautiful bowls and vases.

The five years following the formulation of the Theory of Immune Class Regulation

Less than a year after I made this theory whilst in Mel's laboratory, I decided to move to the Institute of Animal Physiology, situated in the village of Babraham, just outside Cambridge, England. Unfortunately, I did not find this to be a congenial place of work. I soon had my eye on adverts for positions elsewhere. I found one that requested applications from individuals with exactly my interests. This seemed too good to be true! It was for a position at the John Curtin School for Medical Research at the Australian National University in Canberra. Moreover, the head of the department, Gordon Ada, was a well-known immunologist, whom I had met when he visited the Salk Institute. Naturally, I applied for the position. I received a quick response from Gordon, who straight out told me it was a phony advert, as it had been specially crafted with Chris Parish's appointment in mind. No wonder the fit of the requested interests of the applicant fitted mine so perfectly! Nevertheless, Gordon told me, the members of the department were very interested in my application, and that they could support me for two or three years if I wished to come. I accepted eagerly. I had returned to England for less than a year. Although I had started doing some experiments in Mel's lab, I had not produced observations that had led to any publication. I really wanted to become an experimentalist, and felt the John Curtin School would provide me with both a stimulating environment and this opportunity.

Before talking about this new adventure, I have to describe a few pertinent developments in the field. I was naturally eagerly scanning the immunology journals to see whether new reports shed any light on the theory, and whether the theory shed any light on new observations.

Some of my colleagues at the Salk Institute examined the susceptibility of target cells to different forms of attack when bearing different levels of MHC antigens on their surface. They showed that the targets with the lowest amount of antigen were susceptible to antigen-specific CTL, an expression of cell-mediated immunity, but not to antigen-specific IgG antibody-dependent, complement-mediated lysis, whereas target cells with a greater density of antigen were susceptible to both mechanisms.[72] These findings provide clear support for the idea that less foreign targets were susceptible to cell-mediated but not to antibody-dependent attack.

More reports began to appear demonstrating that unresponsive states, for the production of antibody, could be generated in diverse systems upon giving antigen to immunocompetent animals. A remarkable situation developed in time; by taking a bird's eye view of the literature, it was possible to classify these unresponsive states as being generated under one of four different circumstances.[69,73]

i. The administration of high doses of antigen, supraimmunogenic for the induction of antibody.

ii. The administration of low doses of antigen, subimmunogenic for the induction of antibody.

iii. The administration of simple antigens with few foreign sites, such as simple copolymers of the two amino acids, glutamic acid and tyrosine (GT), to certain strains of mice. These molecules were by themselves not immunogenic for the production of antibody, but could induce antibodies when administered complexed to an immunogenic carrier. The generation of this anti-GT antibody response could be prevented in some strains of mice by administering GT before the GT-carrier challenge, an unresponsiveness shown in time to be associated with the presence of GT-specific TsAb cells.[74]

iv. The co-administration of drugs that kill dividing cells, such as **cyclophosphamide**, together with a challenge of antigen that, when administered alone, induced a brisk antibody response.[75] Such drugs were on this basis designated as being immunosuppressive. We shall later discuss the potential pertinence of this activity, of cyclophosphamide and similar drugs, in animal models of cancer therapy.

What was really exciting was that a bird's eye view also showed that there were four general circumstances under which DTH could be induced. These were the same four conditions! However, no one else, other than Chris Parish with his demonstration of low-zone and high-zone cell-mediated immune deviation, had suggested that these unresponsive states for the production of antibody were all instances of cell-mediated immune deviation. Moreover, all these four conditions were explicable on the Threshold Hypothesis. In the case of immunosuppressive drugs, it was clear that their administration might act to partially kill dividing CD4 T cells, and so a situation that led to robust CD4 T cell cooperation and antibody production in the absence of the drug might well lead to tenuous CD4 T cell cooperation in the presence of the drug and hence to the generation of T_{DTH} cells. There was indeed a report that the administration of cyclophosphamide resulted in the modulation of a humoral response into a DTH mode.[76] Many new observations fitted into the conceptual scheme of the theory. I became ever more convinced that there was some validity in how the theory accounted for all these observations. A major impetus for writing this book is to engage my colleagues in discussing this theory and exploring its potential implications in preventing and treating clinical situations.

Australia and new friends

I was lucky in my early scientific life to find myself in institutes established with an exceptional spirit of largesse. The Australian National University (ANU) was another of these. It was established by an act of the Australian Parliament in 1947. Attractive terms were made to both attract Australians back to their home country, and to encourage foreigners to visit the university. For example, I visited as a senior postdoctoral fellow. The airfare from England to Australia for both me and my wife was paid for. In addition, when we left, our airfare to Canada was also paid for and there was a substantial allowance to transport our belongings.

I loved my time in Australia. Two aspects of life there made an indelible impression on me. I found the countryside and the animals extraordinarily beautiful. This is perhaps because they were so different from what I was used to and so I could see freshly, as a child sees the world. I experienced the visual magic of the place on a semi-continuous basis.

Second, I think there is, for many English ex-patriots, a feeling of liberation when going to another English speaking country where there is not the kind of class system that is in place in England. No one could call the Australians stuffy. We had many social interactions at which people took time to discuss all sorts of things, including science. We had good times together, eating, drinking, and conversing.

I became friends with Alistair Cunningham and got a bit of space in his lab to work. He had been the first graduate student of Kevin Lafferty, both of whom will reappear later in my story. Kevin was in another department of the John Curtin School of Medical Research, but would occasionally stroll over and ask if anyone wanted to go out for an ice cream, or sometimes for a beer. Much time was spent chatting over the daily coffee and tea, often on scientific topics. In this sense, the John Curtin was much like the MRC Laboratory of Molecular Biology in Cambridge. This type of collegial interaction was rare in those days, and more so today. People are just too busy to chat about science, foreclosing the possibility that a new vista may develop. Alastair and Kevin obviously had a good relationship and mutual respect, and both shared the gift of enjoying life.

I met JFAP Miller, who had made important contributions to immunology as already outlined, at a conference shortly after I arrived. Jacques appeared to me to be very intense indeed. I could not help but see him as a kangaroo, never taking off but always appearing just about ready to make a big jump. He offered me his view that it was difficult to test the Two Signal Model, which made it somewhat meaningless. I met him some twenty years later at the Eliza Hall Institute in Melbourne and was overwhelmed by his unexpected charm and kindness. I wanted advice on how to thymectomise mice. He told me there was only one way to learn. He thereupon took me into the animal quarters and showed me every detail of the art. It is a sophisticated and difficult operation. I made copious and detailed notes on the methodology as I flew out of Melbourne, and we managed to establish the technique in my lab. However, we never actually used it in substantial experiments.

There were two immunovirologists in the Microbiology Department of the John Curtin, Bob Blanden and Arno Mullbacher. I never formally collaborated with either of them, but they were people I enjoyed discussing science with, and much else besides. Bob had broad interests and followed them wherever they led. He also had the fearless courage to take stances that he believed in, even if it was not politically the most expedient action to take. He interacted a lot with Arno Mullbacher who was from Austria and is married to a charming Australian lady named Penney. I imagine that Austria might compete successfully with England for the strictness of its social mores. Arno has an unkempt beard and intense blue eyes. He looks as though he must be a bit of an anarchist. I feel we have this in common, probably due to similar reactions to aspects of our heritage. When I experience the rigidity of social structure, it makes me wish to shake things up.

I am warm about many of the people I met in Australia, as I felt they were individualistic, treasuring their independence. I also found this individualism associated with a tolerance for eccentric behavior on the part of others. It was a comfortable and inspiring atmosphere. This impression may reflect the fact that most people I met were academics.

Going to Canberra

Arriving in Canberra in 1975 was exhilarating. Excitement was in the air. Peter Doherty and Rolf Zinkernagel had recently done their first experiments concerning the nature of the recognition of virally infected cells by cytotoxic T lymphocytes. It is probably fair to say that Bob Blanden locally established a lot of the methodology, and even some of the experimental systems, that allowed Peter Doherty and Rolf Zinkernagel to make these observations on cytotoxic T lymphocytes, for which they received their Nobel Prize. Naturally, I was eager to meet Chris Parish and chat with him about our different ideas. Alastair Cunningham and Kevin Lafferty had been trying to understand the basis of graft rejection, and had come to the conclusion that the activation of T cells required two signals. This was somewhat in line with what we had proposed. Their signal 1 and our signal 1 were identical, but their signal 2 was suspiciously like, but critically different, from ours, though they gave us credit for postulating a requirement for a second signal. The Lafferty/Cunningham model of T cell activation became prominent and important in the field. Alastair Cunningham and Linda Pilarski were also trying to demonstrate that the stimulation of B cells could result in progeny with mutations in the genes coding for their Ig receptor, resulting in changes in the specificity of the receptor. This was a hypothesis I had independently come to favor. I found bench space in Alastair's laboratory. All this research was of considerable significance, as will become apparent. However, I would like first to describe a real experimental collaboration that led to what we felt were interesting and beautiful results.

Testing some predictions of the Theory of Immune Class Regulation

Chris Parish had an English graduate student, Ian Ramshaw. Ian was clearly a bright young man. He became enthusiastic about my Theory of Immune Class Regulation, and we decided to try to work and plan experiments together to critically test some of its main predictions. We chose to test whether the theory correctly described the basis of humoral and cell-mediated immune deviation at the cellular level, with the idea that the two of us would do experiments in parallel. The theory stated that there were T_sAb cells in an animal expressing DTH, and whose immune response was locked into a cell-mediated mode. No one had shown at that time that these three attributes of an immune state were concurrently expressed, as we anticipated. We explored the literature to find a suitable system. In addition, we decided to test whether a state of humoral immune deviation had three corresponding attributes: production of antibody, lack of expression of DTH, and the presence of T_sDTH cells, that is T cells that would suppress the induction of a DTH response. We worked closely together every day, discussing and exploring how we might realize our objectives. Ian often led the way and found a paper that led him to suggest an experimental approach to realize our aims.

The time came when I had to think about another position. I decided to do a big trip across the US and Canada. Linda Pilarski, who had worked with Alastair Cunningham as a postdoctoral fellow in Canberra, had gone to Edmonton, Canada, and helped to arrange the Canadian portion of the trip. In the end, I decided that when I left the Australian National University, I would go to Edmonton.

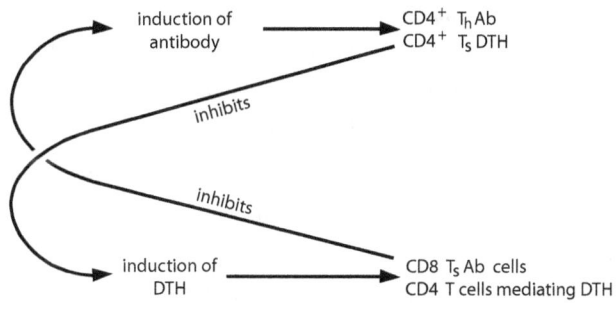

Fig. 24. Summary of the findings demonstrating the existence of TsAb and TsDTH cells

I was absolutely flabbergasted when I got back. Ian's experiments had gone exceptionally well and the results were beautiful. I shall describe later some conclusions his observations led to. However, the gist of Ian's findings supported the proposals of the theory that are summarized in Figure 22. Ian showed that there are T_sAb cells in mice in a state of cell-mediated immune deviation, and these T_sAb cells were CD4-CD8+ T cells, referred to simply as CD8 T cells,[77, 78] see Figure 24. In addition, Ian could demonstrate the existence of T_sDTH cells in mice immunized to be in a state of humoral immune deviation; the immuninized mice produced antibody, expressed no DTH reactivity, and were resistant to the induction of DTH. These T_sDTH cells were, in contrast, CD4+CD8- T cells, referred to simply as CD4 T cells,[78-79], see Figure 24. Ian also examined the markers on T_{DTH} cells, i.e. those cells able to mediate DTH. He found they were also CD4 T cells.[78] We had shown by these experiments that there are two different types of CD4 T cell that differ in three respects. One, the T_{DTH} cell, mediates DTH, is generated under conditions where antigen induces a DTH response, and this T cell does *not* suppress the generation of DTH responses. The other CD4 T cell, the T_sDTH cell, does not express DTH, is generated under conditions leading to a vigorous antibody response, and has the ability to inhibit, i.e. suppress, DTH responses. Ian and I were excited about our successful testing of some critical ideas of the theory, and we published our results with Chris Parish in good journals.[77-79] The impact we anticipated did not materialize. The conclusions these experiments led to are summarized in Figure 24. They verified and gave substance to some of the predictions of the Theory of Immune Class Regulation depicted in Figure 22, and provide further information on the characteristics of the critical T cells involved in suppressing these opposing classes of immunity.

Synopsis of Chapter 4

I became interested in 1971 in how antigen interacts with cells of the immune system to determine the class of immunity induced. The Theory of Immune Class Regulation I published in 1973 has not changed in essence in the last forty years. I summarize three aspects of the theory in this synopsis: ideas on the biological significance of immune class regulation, proposals as to how the exclusivity between the generation of cell-mediated and humoral responses is achieved, and a proposal for the mechanistic basis of the decision criterion controlling whether antigen predominantly generates cell-mediated or antibody responses.

*Observations show that autoimmunity against a self antigen can sometimes be induced when a foreign antigen, which cross-reacts with this self antigen, impinges upon the immune system. Such findings are anticipated in the context of the Two Signal Model of lymphocyte activation. As the impingement of such crossreacting antigens is anticipated to be frequent, it is natural to consider whether immunological autoreactivity is often induced. Sensitive assays reveal that low levels of autoantibodies can often be detected in apparently healthy people. This finding has led to a recognition of the importance of distinguishing between immune **autoreactivity**, that may be benign and have no pathological consequences, and **autoimmunity**, a form of autoreactivity that by definition is pathological, being associated with destruction of host tissues, cells or molecules.*

An examination of how IgG antibodies activate effector functions, such as IgG-dependent, complement-mediated lysis of cells,[66] or IgG antibody-dependent NK cell-mediated lysis,[80] shows that such mechanisms will only be effective against cells with at least several hundred thousand sites recognizable by antibody. Such findings can explain why the presence of autoreactive IgG antibodies is often not pathological. Thus cells with less that the required number of antibody-recognizable sites will not be attacked by IgG antibody-dependent mechanisms. I surmised such cells would generate an effective cell-mediated response.

Independently of these observations, Pearson and Raffel had proposed, again on the basis of observation, that minimally foreign cells and antigens can only induce cell-mediated responses. This made sense if IgG antibody, even if produced, was ineffective. It seems plausible that there are three rules governing the regulation of the class of immunity an antigen induces; these rules make both biological sense and are supported by observation. First, minimally foreign antigens only induce a cell-mediated response, as only a cell-mediated response can be effective. Consider now immune responses to more foreign antigens. The second rule is that an effective immune response should be generated shortly after infection of an animal or shortly after antigen impacts upon the immune system. Given the quantitative features of the IgG-dependent effector functions described above, low levels of IgG antibodies cannot be effective in mediating attacks on an invader. It is not possible to produce substantial levels of IgG antibody shortly after infection/antigen impact, in the absense of sufficient time for sufficient B cell expansion. Thus antibody, even if produced, would be ineffective and so an effective cell-mediated response is generated instead. The price paid is that any autoimmunity generated is more damaging than a corresponding antibody response would be. The third rule is that the response should not be more "vicious" than necessary, as this would only increase the debilitating consequences of any autoimmunity generated. This rule accounts for why the immune response against a foreign antigen evolves into a predominant antibody mode once the potential to produce larger and effective levels of antibody has been reached; this phase of the immune response is associated with the suppression of the corresponding cell-mediated response. This inhibition of cell-mediated immunity minimizes the damage caused by any autoreactivity generated. These three rules account for why cell-mediated and antibody responses are regulated in the observed fashion. I suggest such regulation arose so that immune responses against foreign invaders are generally effective whilst, at the same time, minimizing the damage of any autoreactivity generated.

An examination of immune responses in intact animals shows there is a tendency for the response to be either of a cell-mediated or an antibody phenotype. This tendency for exclusivity is most clearly seen in situations where the immune response to a particular antigen has been locked into a cell-mediated or a humoral mode. Such locked states are referred to as states of cell-mediated and humoral immune deviation. I proposed in my Theory of Immune Class Regulation that exclusivity between classes of immunity was due to the generation of T cells, associated with one mode, that inhibit the activation of lymphocytes involved in the opposing mode, see Figure 22. This idea was strongly supported by experiments carried out by Ian Ramshaw, a graduate student of Chris Parish. He showed that cell-mediated immune deviation was associated with the presence of antigen-specific CD8 T cells that can inhibit the generation of antibody responses, referred to as TsAb. He also showed that humoral immune deviation is associated with the presence of *antigen-specific CD4 T cells that inhibit cell-mediated responses in the form of delayed-type hypersensitivity (DTH), referred to as TsDTH cells. Ian's findings are summarized in Figure 24. He also showed that cells mediating DTH were CD4 T cells. The characterization of these two different classes of CD4 T cell, with quite distinct biological activities, preceded the subsequent discovery that clones of CD4 T cells can be classified into two types, reflecting the existence of Th1 and Th2 cells, by over a decade. We described the discovery of Th1 and Th2 cells in the next chapter.*

A knowledge of the mechanisms underlying cell-mediated and humoral immune deviation do not address what controls whether antigen generates, in the first instance, a cell-mediated or an antibody response. The Threshold Hypothesis proposes how antigen interacts with the cells of the immune system to favor the generation of one class or the other. This hypothesis was based upon the then tentative postulate that the activation of CD4 T cells requires antigen mediated CD4 T cell cooperation, which is now supported by diverse observations. The Threshold Hypothesis states that tenuous CD4 T cell interactions lead to the generation of Th1 cells and cell-mediated immunity, whereas robust interactions are required for the generation of Th2 cells and antibody responses.

This hypothesis was proposed because it accounted for all the variables of immunization then known to affect the cell-mediated/humoral nature of the ensuing response. In particular, it accounted for why minimally foreign antigens could only induce cell-mediated immunity, as there will be relatively few CD4 T cells specific for such antigens, and so CD4 T cell cooperation will be tenuous. There are more CD4 T cells specific for a more foreign antigen. Consider what happens if an optimal amount of antigen is administered that supports robust CD4 T cell interactions shortly after the antigen is administered. An antibody response ensues. Now consider what would happen as the dose of antigen is successively lowered. The strength of the antigen-mediated CD4 T cell interactions will decrease until they are tenuous, and a cell-mediated response is generated. Now consider further what will happens as antigen stimulates CD4 T cells to divide and so the CD4 T cells become more numerous. So long as the antigen level is sufficiently sustained, the strength of the CD4 T cell interactions will increase, and the response will evolve with time into an antibody mode. Thus, the Threshold Hypothesis accounts for how the nature of the immune response depends upon the degree of foreignness of the antigen, on the dose of very foreign antigens and the time after immunization that its nature is assessed, and as characterized by Salvin (Figure 8 of Chapter 1).

One striking prediction can be made on The Threshold Hypothesis. Suppose we have a situation in an intact animal where antigen generates a brisk humoral response. Suppose we keep all variables constant except we employ a means for reducing the number of CD4 T cells in the animal. If the reduction is of an appropriate amount, the reduction will decrease the strength of CD4 T cell interactions, so that a cell-mediated response is now generated instead of the brisk antibody response. We shall see that this prediction has been verified in a number of different systems.

Chapter 5

MAJOR DEVELOPMENTS IN THE
FIELD FROM 1970 TO TODAY

These developments are of great interest for their own sake. Their description will also provide a context for my proposal of an integrated view of how the immune system functions, and how such a view can be employed to envisage and devise strategies to prevent and to treat various clinical situations of interest.

The major histocompatibility complex (MHC) and MHC molecules

One of the most exciting constellation of immunological events of the last fifty years has been the revelation of the central roles of MHC molecules in diverse immune processes. This new knowledge impinges upon virtually all aspects of the immune system. As just one example, this knowledge has posed problems for the models of B cell and T cell activation that Cohn and I proposed in 1970. We trace here how the roles of MHC molecules came to be recognized and were elucidated.

Medawar had come across many burn victims during WWII and was inspired to become a pioneer of transplantation biology. He hoped to treat burn victims by transplanting skin from willing donors. This aspiration was not realized, but his research contributed to the development of transplantation immunology and the discovery of the genetic region known as the *major histocompatibility complex (MHC).*

Grafts, for example skin grafts, between mice of different strains are almost always rejected. This is known to be due to an immune response against incompatible, i.e. foreign, *histocompatibility antigens* present on the surface of the cells of the graft. That rejection is due to an immune response, rather than to mechanisms of innate defense, is supported by two kinds of observation. First, the rejection of a second graft, obtained from a genetically identical donor as the initial graft, is rejected more rapidly than the initial graft. This memory is specific, and is an expression of a secondary response, called second set rejection. In addition, Billingham, Brent, and Medawar gave newborn mice of one strain a mixture of cells from another strain, and as adults they were tolerant, or partially tolerant, of the donor antigens, as assessed by their acceptance or partial acceptance of donor skin grafts. These were the experiments for which Medawar received the Nobel Prize. These observations were interpreted as demonstrating that the rejection of histoincompatible grafts by adults could be prevented if the adult had been neonatally exposed to the antigens present on the graft, and so learnt to regard the antigens as "self antigens".

Transplantation studies also first revealed the existence of a genetic region, which occurs in all vertebrates, called the major histocompatibility complex. This genetic complex has had the habit of declaring its importance several times in the history of immunology.

Grafts between different strains of mice are rejected at different tempos, depending upon the pair of strains involved. It appears that primary grafts are either rejected in fewer than fourteen days or in more than twenty. It was found by genetic analysis that mice that mutually reject primary grafts in less than fourteen days are genetically different in a certain region called the **major histocompatibility complex (MHC)**. In fact, there are many different histocompatability loci coding for histocompatibility antigens; however, the tempo of graft rejection is fast if there are MHC incompatibilities between the donor of the graft and the host and so, in this sense, differences in the MHC region are dominant in determining the speed of rejection.

The name MHC wisely includes the word 'complex' because genetic and molecular studies reveal that it consists of several closely linked genetic loci. A genetic locus is a site in the genome, or at a particular stretch of the DNA of a chromosome, where different **alleles** or alternative genes can exist, either in the two homologous, parental chromosomes that an individual has, or in different individuals of the species. Normally, there are only a few alternative alleles at one genetic locus, and they typically differ minimally from one another, affecting only one or two amino acids of the polypeptide chains they encode. In the mouse, there are six important loci in the MHC region. These all share two remarkable features. First, there are many alleles found in different individuals at any one locus, typically at least forty. In humans there can be more than a hundred alleles at one of these genetic loci. Second, when any two alleles associated with the same locus are compared, it is found that they differ greatly, affecting many amino acids in the polypeptide chains they encode, typically at least twenty. The six important MHC loci in the mouse are all close together on one chromosome, and so are inherited as a block about 99% of the time. Thus, intra-MHC recombination occurs rarely.

The MHC is the most polymorphic genetic region known in biology. What is the significance of this polymorphism and how is it related to the function of MHC molecules? These are questions we shall address.

There are two classes of MHC molecules, known as class I and class II. Class I MHC molecules exist on the surface of all cells. Class II occurs only on certain cells, such as macrophages and other phagocytic cells. The functions of class I and class II MHC molecules are distinct but also have similarities. It is convenient if we first consider the function of class I MHC molecules.

The function of class I MHC molecules and the specificity of CTL

The class I MHC molecules constitute a central part of a spy or immune surveillance mechanism that the immune system uses to monitor what is going on inside a cell. To set the scene, consider an intracellular parasite that tries to partially take over host machinery for its replication, but does not allow the insertion of any of its own molecules into the external membrane of the infected cell. The immune system would appear at first glance to have no means of distinguishing the infected from an uninfected cell, and would therefore be unable to protect against this invader. Consider another scenario. Suppose some mutations occur in a cell that affect the function of internal proteins that control the cell cycle; the mutated cell grows in an unrestrained fashion and is cancerous. Again, it would appear that the immune system has no way of distinguishing this cell from a benign cell. These examples are, perhaps, somewhat fanciful. However, they provide instances that allow one to imagine and recognize the potential importance, for the functioning of the organism, that the immune system has the ability to detect what is going on inside a cell. It appears that class I MHC molecules play a pivotal role in this surveillance.

What appears to happen is this: samples of proteins synthesized inside a cell are subjected to degradation in a protein complex called the **proteasome.** The polypeptide chains of most typical proteins contain at least a hundred amino acids linked together. The proteasomes cut the polypeptide chains into pieces that range in size, but typical fragments are about 10-20 amino acids long, and are not expected to have a stable, three dimensional structure as the proteins from which they are derived. The class I MHC molecules contain a groove at one end of the molecule, into which some peptide fragments bind. Grooves of all class I MHC molecules have an affinity for the common backbone of the oligopeptide, but there are amino acid residues that must fit pockets or avoid protuberances of the groove for optimal binding. Thus, a given MHC molecule selects certain peptides able to bind to its groove. The loaded MHC molecule is then transported to the surface membrane of the cell, where its outward-facing tip presents the peptide-bound groove to the world of lymphocytes. In this way, the immune system monitors any changes in the internal proteins of the cell. Moreover, much of the variation found between class I MHC molecules is concentrated in and around the groove. Thus different class I MHC molecules in general bind to and present different peptides.

We need to know more about the genetics and chemistry of class I MHC molecules if we are to truly appreciate their biological role. Inbred mice inherit identical chromosomes from each parent, unless they are male, in which case they receive an X chromosome from their mother and a Y chromosome from their father. However, the MHC is not on the sex chromosomes, so both males and females of inbred strains are homozygous for the MHC.

Class I MHC molecules consist of two polypeptide chains, one of which does not significantly vary within the species and is called β2 microglobulin. Mice have two genetic loci, called the K and the D loci, and the genes at both these loci code for the highly variable chain of different class I MHC molecules. There are many different possible alleles or alternative genes at the K locus, and these are designated as K^d, K^q, K^s, K^k, etc. Similarly, there are many potential alleles at the D locus, referred to as D^d, D^k, etc. Mice of an inbred strain with a K^a and a D^a allele will have two kinds of MHC class I molecules, those having the polypeptide chain encoded by the K^a allele, assembled together with β2 microglobulin, and the other having the polypeptide chain coded for by the D^a allele, together with β2 microglobulin. We refer to these as K^a and D^a class I MHC molecules. As mice of inbred strains have identical MHC chromosomes, their cells express just two class I MHC molecules, their K and D molecules. The MHC chromosomes that wild mice mice inherit from their two parents usually carry different K and D alleles, so the cells of wild type mice have four different class I MHC molecules, two different K and two different D molecules.

The studies of Peter Doherty and Rolf Zinkernagel represent the first observations that led in time to this picture of how class I MHC molecules function.[81] They infected mice with a particular virus, and harvested splenic lymphocytes from the mouse a few days later. They were able to detect amongst these cells *cytotoxic T lymphocytes (CTL)* able to lyse host cells infected with that virus, but unable to lyse uninfected host cells, or host cells infected with unrelated viruses. These CTL were thus virus specific. However, Peter Doherty and Rolf Zinkernagel made the cardinal observation that these CTL did not lyse other target cells infected with the same virus unless the target cell came from a mouse that shares one or more MHC class I molecules with the infected mouse. Optimum lysis usually required the target cell to have the same alleles at both the K and D loci as the strain that produced the CTL. These observations are summarized in Figure 25. They were interpreted to mean that the CTL, in order to lyse the target cell, had to recognize both a viral product and a part of the class I MHC molecule that differed between class I MHC molecules. We now know in molecular detail how this occurs, as the structure of the T cell receptor and class I MHC molecules have been solved in atomic detail by protein Xray crystallography. The T cell receptor has a broad binding site that recognizes not only a viral peptide bound to the groove of the class I MHC molecule, but also the lips of the groove of the MHC molecule. So the T cell receptor recognizes the peptide/ MHC complex.

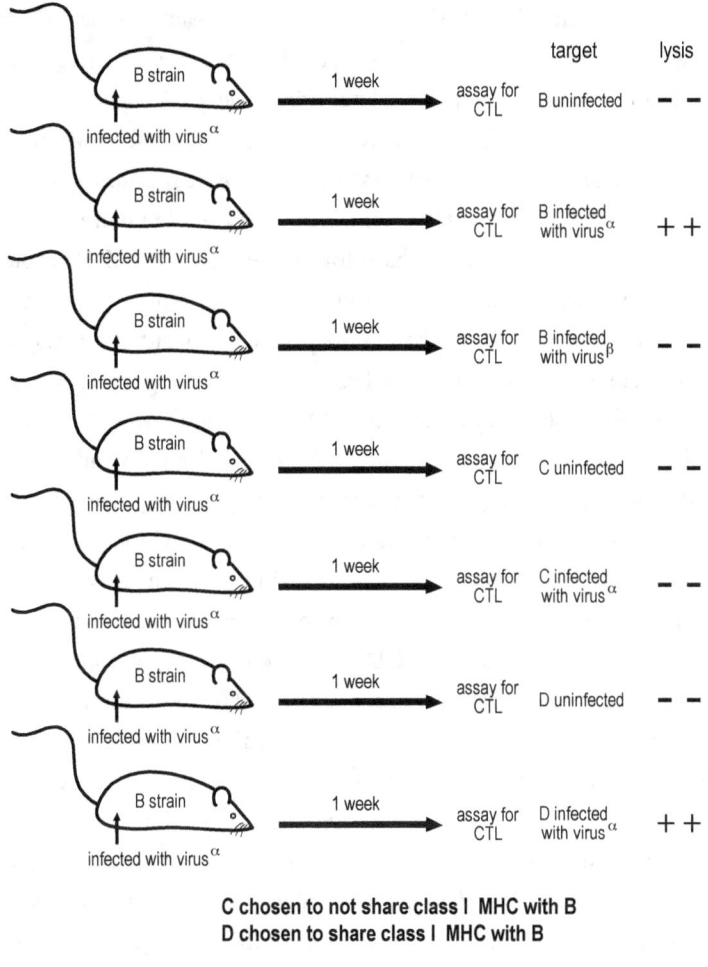

C chosen to not share class I MHC with B
D chosen to share class I MHC with B

Fig. 25. CTL and target cells must be class I MHC matched in order to obtain target cell lysis

An interesting experiment vividly illustrates the nature of this surveillance system. It is possible to raise CTL that recognize influenza virus-infected cells. One of the viral proteins is called the *nuclear protein,* or *NP,* because it was originally only readily detectable in the nucleus of infected cells. Observations provided indirect evidence that some of the influenza-specific CTL were recognizing the nuclear protein, hardly understandable on the old view that this protein was only readily found in the nucleus of infected cells. The discovery was made that *uninfected cells* could become targets of these NP-specific CTL if these cells were incubated with certain oligo-peptides, about ten amino acids long, that were derived from the NP. A few of the class I MHC molecules on the surface of the target cells lose the peptide that is bound to their groove, and so their empty grooves are able to bind the NP-derived peptide. "This peptide decoration of the target cell from the outside" now allows the NP-specific CTL to bind to the decorated target, leading to its lysis.

Does this insight into how CTL recognize antigen lead to ideas on why MHC molecules are so polymorphic? It turns out that there are rare instances where a class I MHC molecule does not bind any viral peptides when a certain virus infects a cell. This means that this particular class I MHC molecule is useless, when this cell is infected by this virus, as a reporter of viral infection to the immune system. Luckily, inbred mice have two class I MHC molecules, so a double failure would be rare. Wild type mice would most usually have four different class I MHC molecules, as there are so many alleles at each locus. Thus, a failure to report an infection would be considerably rarer in wild-type than in inbred mice. Consider the imaginary situation where there is only one class I MHC molecule among mice. A mouse virus whose peptides did not bind to this class I MHC molecule would cause havoc, as there would be no CTL generated, and so the infected mice would likely all die, assuming the virus is pathological. It appears that the polymorphism of class I MHC molecules, and the existence of more than one class I MHC locus, ensures that an intracellular pathogen cannot succeed in realizing such an evasion mechanism, effective throughout a population of mice, or of humans for that matter. The major conclusion of all these studies is that CD8 CTL do not recognize antigen itself, but rather an antigenic peptide in the context of class I MHC molecules. Indeed, protein crystallography has allowed investigators to determine the molecular structure of particular class I MHC molecules binding a particular peptide. This reveals that the peptide fits snugly into the groove of the MHC molecule. Similar studies involving T cell receptors show that the T cell receptor makes contact with both the peptide and the lips of the groove. This detailed molecular picture of how the T cell receptor interacts with its peptide/MHC ligand is most gratifying, as it allows an accounting of many observations, made at the cellular level and at the level of the system, on MHC-restriction.

The function of class II MHC molecules in antigen presentation and CD4 T cell activation

An investigation of the requirements to activate CD4+ Th cells was initially more difficult than investigating the requirements to activate B cells, partly because there was no recognized, simple assay to measure or to detect their function. The lymphocytes specific for a given antigen are scarce in a naïve mouse. If one cultures naïve lymphocytes with an antigen, a scarce antigen-specific CD4 T cell might be stimulated to divide, but the scarcity of the responding cell makes its division difficult if not impossible to detect. When immunologists cultured lymphocytes with antigen from a mouse that had been primed to the antigen, they could readily detect the antigen-dependent proliferation of the sensitized lymphocytes. This general finding supported basic ideas of the Clonal Selection Theory. Memory responses are envisaged to be due, at least in part, to the higher frequency of specific lymphocytes following priming. Moreover, experiments show that this cellular proliferation was critically dependent upon CD4 T cells. This antigen-dependent cellular proliferation provided a simple assay with which immunologists could investigate how CD4 T cells recognize antigen.

It is helpful at this point to anticapte some of the conclusions to be made from observations we shall shortly discuss, and set them into an evolutionary context. This allows us to more readily make sense of the observations.

We have seen a major means of innate defense, present in the simplest invertebrates as well as in vertebrates, is the opsonization of invaders by phagocytic cells. This oposonization is followed by the "digestion" of the invaders by chemical conditions and the action of enzymes, which breaks down the invader's large molecules. These processes, as already noted, result in destruction of the invader and provide food for the host in the form of small molecules out of which host molecules can be made. It might be correctly thought that through evolutionary time the phagocytic cells of non-vertebrates have acquired very efficient means of opsonizing and destroying foreign invaders, and so they might have a central role in initiating immune responses, as they first occurred in vertebrates. This viewpoint became more attractive as the processes involved in the activation of T cells were revealed.

A first set of findings made with such systems was puzzling. It was known that antibody recognizes intact proteins much better than a protein digest obtained by exposing the antigen to proteases that cut up the protein antigen into rather small pieces. One protein used in many studies was **ovalbumin** or **OVA**, the major component of egg white. It was found that OVA-primed CD4 T cells proliferated when cultured with OVA, but they also proliferated equally well in response to a protease-digest of OVA. This type of observation implied that CD4 T cells recognize antigen in a very different way from the way antibody does, and hence very different from the way B cells recognize antigen.

Fortunately, another avenue of investigation, encapsulated in Figure 26, led to more direct insights as to how T cells recognize antigen. It was discovered that purified, primed CD4 T cells did not proliferate in response to antigen, but that another cell, initially identified as a macrophage, was required. Moreover, it was possible to give the macrophages antigen, wait a few hours, attempt to wash away any free antigen, and then combine the macrophages with purified, primed T cells, resulting in the proliferation of the T cells. The ability of such **antigen-pulsed macrophages** to cause T cell proliferation was not abrogated by treating the macrophages with certain poisons, such as formalin, if this treatment took place six hours after the macrophages had ingested the antigen, see group 4 of Figure 26. However, treatment of the macrophages with formalin, either before their ingestion of the antigen or within an hour of ingestion, blocked the macrophages' ability to stimulate purified and primed T cells, see group 5. These important studies led to the recognition that macrophages perform two functions: the first took rather less than six hours after exposure to antigen to complete, and is prevented by treating the macrophages with formalin; the second macrophage function is distinguished from the first by being insensitive to formalin treatment. A critical experiment now showed that formalin-treated macrophages, exposed to a protease digest of OVA, constituting an assortment of peptides derived from OVA, is able to stimulate OVA-primed T cells, see group 6.[82] It seemed probable that there were molecules on the surface of formalin-treated macrophages that could bind OVA-peptides. These molecules turned out to be class II MHC molecules![82]

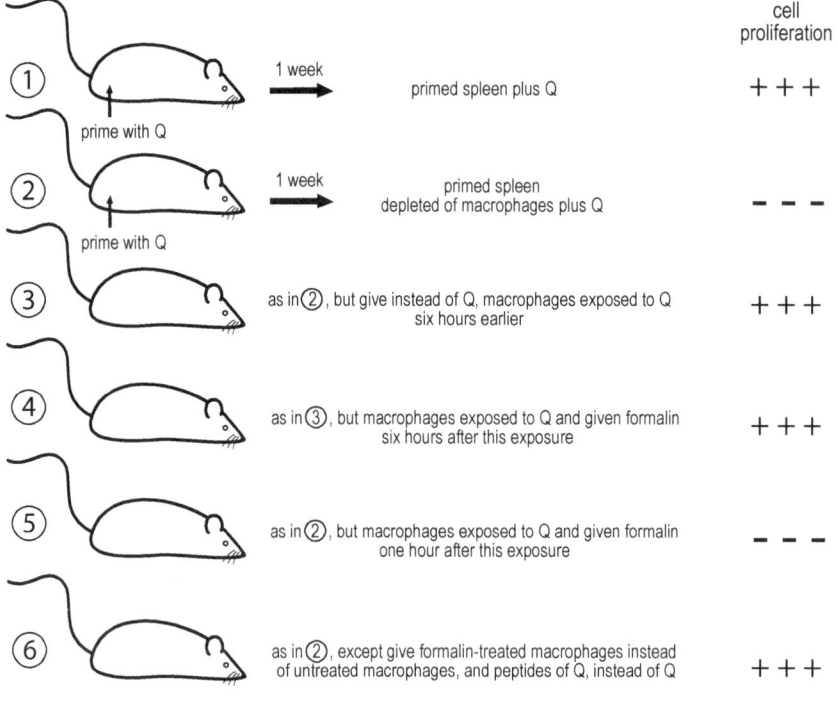

Fig. 26. Observations demonstrating the two functions of antigen presenting cells: antigen processing and antigen presentation.

The first function of macrophages, sensitive to formalin treatment, appears to result in the ingestion of OVA and, after a few hours, the appearance of OVA-peptides on the surface of the macrophage. These events are collectively referred to as ***antigen processing***. We now know that processing involves the uptake of antigen by the macrophage by some process such as opsonization. The antigen lands up in a membrane-bounded compartment of the macrophage called a *phagosome*. This phagosome fuses with another vesicular compartment called a *lysosome* that contains many enzymes. The antigen is thus degraded in this compartment, known as the *phagolysozome*, into oligopeptides of about 10-20 amino acids in length. Newly synthesized class II MHC molecules are directed to this compartment, and these nascent class II molecules pick up those peptides that can bind to their groove. The class II MHC molecules are then inserted into the external membrane of the macrophage.

Macrophages bearing OVA-peptides on their surface are able to stimulate the OVA-specific CD4 T cells to divide. This stimulation could be blocked by adding antibodies that bind to the class II MHC molecules expressed by the macrophages. These observations support the concept that the CD4 T cell receptor binds to a peptide, derived from the antigen, and itself bound to the groove of class II MHC molecules.

This second function of macrophages is called *antigen presentation*. We shall see macrophages are not the only cell type that has this ability. However, all such cells have on their surface class II MHC molecules. Such cells, able to present antigen, are collectively called *antigen presenting cells*, or *APC.*

Critical aspects of this picture appear to have been unambiguously demonstrated. Thus, it has been possible to examine which peptides of an antigen can be presented to CD4 T cells by an APC and to show that this ability shows a strict correlation with the peptide's ability to bind to the groove of the APC's class II MHC molecules. Such studies appear to provide overwheming evidence that class II MHC molecules are the presenting molecules, and lay bare the reason why the nature of the MHC molecules genetically controls responsiveness to antigens.[83]

The evolution of our understanding of the processes of antigen processing and presentation by APC have become almost as central and unquestionable to the field as the tenets of the Clonal Selection Theory. It appears that they represent irreversible progress. Such understanding will most probably not be undermined by future progress but be incorporated into a broader picture.

The MHC-restricted model of the B cell/T helper cell interaction

The realization that Th cells do not recognize intact antigen, but rather peptides derived from the antigen and bound to class II MHC molecules, meant that the antigen-bridge model of the B cell/T helper cell interaction, depicted in Figure 15 of Chapter 3, could no longer be entertained. It is useful here to introduce some terminology that allows a clear discussion of the issues involved. We have often called

Th cells, generated by immunizing with an antigen, OVA let us say, as OVA-specific. The term OVA refers to the intact protein and, when it became clear that the T cell receptor does not bind the intact protein, it was not strictly correct to refer to the Th cells as OVA-specific, even though we shall often do so for convenience. When trying to be more precise, immunologists refer to OVA as the *nominal antigen*. This terminology acknowledges the fact that the T cell receptor does not bind the antigen itself, but rather a peptide derived from the nominal antigen, and itself bound to an MHC molecule. The use of the term nominal antigen can be used in the context of CD4 or CD8 T cells, in which case the protein-derived peptide would respectively bind to class II and class I MHC molecules.

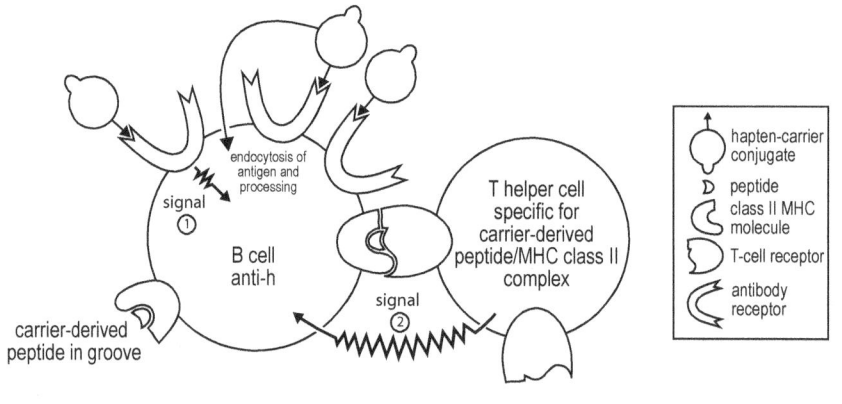

Fig. 27. The MHC-restricted model of B cell/T helper cell interaction of B cell activation

The fact that the receptors of the Th cells do not directly bind the nominal carrier meant that the antigen-bridge model of the B cell/Th cell interaction for B cell activation had to be substantially modified. Antonio Lanzavecchia developed the *MHC-restricted B cell/T helper cell model* of B cell activation.[84] The view he developed was critically dependent on the fact that B cells express class II MHC molecules on their surface. He demonstrated that a hapten specific B cell can bind a hapten carrier conjugate via its Ig receptors, resulting in its uptake into the B cell. The hapten carrier conjugate lands up in an endocytic compartment, where it is processed, ie cut up into oligopeptides of ten to twenty amino acids in length. Newly synthesized class II MHC molecules are directed to this compartment and pick up those olio-peptides able to bind to its grooves. Peptide-loaded class II MHC molecules are tansported to the external membrane of the cell. The hapten-specific B cell can in this fashion present carrier-derived peptides to CD4 T cells specific for the nominal carrier. This presentation would now also allow an activated Th cell, specific for the nominal carrier, to bind to the carrier-derived peptide/class II MHC complex, and deliver signal 2, see Figure 27. Note that the carrier protein will only get into the hapten-specific B cell if the hapten is linked to the carrier. Thus, in order for T helper cells specific for a nominal carrier to be able to help the activation of hapten-specific B cells, the hapten must be physically attached to the carrier. This requirement for linkage is critical as it ensures that the delivery of T cell help is focused only onto B cells specific for the nominal antigen or a hapten linked to this antigen. This requirement is so important we have a specific way of referring to it. We say the interaction between the B cell and the T helper cell requires the *operational recognition of linked epitopes*.

MHC genes control the ability to immunologically respond to antigens

The understanding of the role of class II MHC molecules in antigen presentation accounted for another set of observations. These studies were directed at discovering why some animals can make a response to an antigen and others cannot do so. One of the first examples of this phenomenon was discovered in guinea pigs in their response to *poly-L-lysine (PLL)*. This system was the one I came across in the literature as a graduate student. Non-responder guinea pigs do not respond to a challenge of PLL but do produce anti-PLL antibodies on challenge with PLL conjugated to BSA. Other studies in mice also employed simple and synthetic polypeptides, such as copolymers of *glutamic acid and tyrosine (GT)*, or copolymers of *glutamic acid, alanine, and tyrosine (GAT)*. In the cases of PLL and GAT, it was found that some strains produce antibody and others do not, when given a standard challenge of antigen. Interestingly, it was found that a single genetic locus, or alleles at a few closely linked loci, determine the ability to respond, and that the ability is dominant. This genetic locus was called many years ago the *Immune response-1* or *Ir-1 locus*.

Genetic studies eventually showed that the ability to respond maps to the loci encoding class II MHC chains. Thus, the ability to respond depends on having certain class II MHC molecules. How many and how well the peptides of the nominal antigen bind to the grooves of the class II MHC molecules an individual possesses has a substantial effect on how many Th cells specific for the nominal antigen there are in this individual. If none of the peptides generated bind to the grooves of class II MHC molecules, there will be no CD4 T cells specific for the nominal antigen. There will be few, if any, Th cells in a non-responder strain and more in a responder. The progeny of responder and non-responder mice, belonging to inbred strains and so homozygous, will inherit class II MHC molecules from both parents. The class II MHC molecules inherited from the responder parent will be able to bind peptides of the nominal antigen, thus explaining why the ability to respond tends to be dominant. Thus the studies on the role of APC in the activation of Th cells naturally explains the major facts known as to how Ir-1 genes control the ability to generate antibody responses to such antigens as PLL and GAT.

The GOD of antibody genes

The insight into the role of MHC antigens in diverse immunological processes can only be compared in its impact on the field with that arising from the elucidation of how the genes for the B cell and T cell receptors are generated. I remember I was still in Mel's laboratory when Tonegawa's first paper came out demonstrating that the DNA encoding Ig light chains was in a different configuration in the DNA extracted from antibody-producing cells from that in the DNA extracted from embryonic cells.[85] This paper, and the work that followed, are milestones of the field, for which Tonegawa was awarded the Nobel prize. I have to say that, when Tonegawa's paper was published, I felt it was clearly the end of one era and the beginning of another.

Tonegawa's findings gave substance to Lederberg's remarkable speculations,[34] which turned out to be insights, as to how the generation of the diversity of antibody genes is realized. Lederberg's insights were so remarkable that they are, for me, a cause of wonderment and so of reflection: what can be the basis of such accurate insights that go so far beyond what has been observed? This is a philosophical question I entertain but will not pursue here.

I illustrate the essence of Tonegawa's findings by first outlining how a kappa light chain gene is produced in a mouse. A gene corresponding to a functional kappa light chain consists of two DNA elements that are non-contiguous in embryonic DNA. These two DNA elements are brought together by one DNA-joining event as the B cell develops. These two DNA elements are referred to as the V_k and the J_k elements. There are about a hundred different V_k elements and four J_k elements. Each V_k element can join with a J_K element in several ways, giving rise to considerably greater than $100 \times 4 = 400$ possibilities. The number of kappa chains that could be generated, with other mechanisms of variability not considered here, might be estimated as 12,000. Heavy chains are made up of three DNA elements, the V_h, the D_h and the J_h elements, and so the production of a heavy chain involves two DNA joining events.

These joining events, with about 150 V_h, 12D_h and 4 J_h elements, result in a minimum of $150 \times 12 \times 4 = 27,200$ possible heavy chains and, with other mechanisms of variability not considered here, with an estimated 2,400,000 possibilities in total. Different heavy chains can combine with different light chains. Perhaps only 10% of these fold up nicely together, so this gives us a number of potential viable combinations of light and heavy chains of $1/10 \times 12000 \times 2,400,000$, which is about 3,000,000,000. This calculation may not be accurate; nevertheless, it illustrates how the combinatorial use of a rather few genetic elements can give rise to enormous numbers of antibody genes and of even more distinct antibody molecules. It also gives substance to Lederberg's vision.

A major question arises from the nature of this process: do most Ig gene rearrangements that occur give rise to viable antibody chains? The answer is no. There is an elaborate process to check the nature of the Ig chains that the diverse rearrangements produce, and much discarding of cells that have made unsuccessful attempts at creating genes coding for functional antibody chains. An essential element of this checking process is what happens once a rearrangement is made that appears, by a series of tests, to code for a functional Ig chain. It appears that this success results in signals to stop further attempts at rearrangement. Once one has a good hand, it is better to stick than to twist. This inhibition of further attempts at rearrangements is fundamental. It accounts for why lymphocytes have functional genes for only one light and for only one heavy Ig chain, and so are monspecific. This inhibition of further attempts at rearrangement has been exploited to achieve technically important tools for incisive investigations, as I immediately outline below.

Technological developments critical for progress

Shortly after I became an immunologist in the early 1970s, I remember talking to James Howard, then a patrician of the field. He complained how it had become impossible to keep up with new developments. In his early days, he had known every significant immunologist and what they were up to. Now, in 2015, several thousands of pages of progress are added to the literature every month. This book is an attempt to transcend these circumstances. To achieve this, it is necessary and fascinating to gauge how contributions of different natures are essential to progress, and to see how these advances in different areas impact one upon another. I explain here how certain technological developments have only been possible as a result of the success of basic studies, and been essential, in their turn, in transforming the field. I describe both the basis of these technologies, and how their exploitation resulted in great advances in our understanding of mechanisms.

A major technical difficulty in studying the activation and inactivation of antigen-specific lymphocytes is their extreme scarcity. There are estimated to be around a few hundred CD4 T cells in a mouse specific for a typical foreign protein[86] among its roughly 5×10^7 CD4 T cells. Some technical developments have circumvented the problem of scarcity that arises when trying to study how specific lymphocytes interact with their antigen. These technologies include the generation of transgenic mice, all of whose lymphocytes bear the same antigen receptor, and of clones of T cells and of B cell and T cell hybridomas.

The construction and use of transgenic mice

The remarkable insights into how immunoglobulin and T cell receptor genes are generated has allowed immunologists to breed mice with extra genes, encoding, for example, antigen-specific lymphocyte receptors. As we have just seen, the occurrence of DNA arrangements, leading to the production of functional light and heavy Ig chains in a developing B cell, stops further attempts at rearrangement of the Ig genetic elements, such as the V_h, D_h and J_h elements. Mice can be bred that have extra genes that code for functional light and heavy Ig chains, attached to small regions of DNA that are important in controlling the expression of these genes in developing B cells. These extra genes, together with the DNA elements controlling their expression, are inserted at a random site in the genome. These extra genes are called *transgenes*, and mice having them are called *transgenic mice*. The presence of these Ig transgenes leads to the production of the transgenic antibody in developing B cells and, therefore, in the suppression of the rearrangements of endogenous Ig DNA elements, so every B cell in the body expresses the transgenes and therefore the transgenic Ig antibody. This remarkable technology has been used to study, for example, the inactivation of B cells.

David Nemazee created mice transgenic for an antibody that recognized a foreign class I MHC molecule. All the B cells of such mice express the transgenes, and so had Ig receptors for the pertinent MHC class I molecule. David examined what happened when this transgenic mouse was bred with another mouse so that the pertinent class I MHC molecule was now a self antigen. Such mice have few, if any, B cells![87] Chris Goodnow also generated transgenic mice, but in his case the transgenes coded for an IgM antibody that bound to the antigen **hen egg lysozyme (HEL)**. Chris also generated other mice, transgenic for a gene encoding HEL. The double transgenic mice, mice bred to have both the Ig and HEL transgenes, had a deficiency of functional B cells, as the B cells were specific for the self antigen, HEL, and were rendered non-functional as they were generated.[88] We shall see later how such transgenic systems have been exploited to reveal mechanisms of B cell inactivation.

The discovery and the usefulness of interleukin-2.

Another important means of overcoming the scarcity of lymphocytes is to clone them, i.e. generate a population of lymphocytes derived from one specific lymphocyte. The ability to clone CD4 T cells arose through an analysis of what activated CD4 T helper cells produce when they are stimulated with antigen.

The stimulation of activated T cells with antigen leads to the production and secretion of molecules that affect the proliferation and differentiation of other lymphocytes or white cells. Immunologists envisaged that the production of such molecules reflects the way one antigen-specific lymphocyte can affect the differentiation state of other white cells. One example illustrates the nature of these observations. It was found that OVA-immunized CD4 T helper cells, able to help the activation of B cells, produce, upon stimulation with OVA in the presence of APC, molecules that could in turn stimulate certain cultures of purified B cells to differentiate into antibody producing cells. Molecules mediating such effects have been given different names. One is *interleukin*, a messenger between leucocytes, for a molecule made by one leucocyte (white blood cell) that influences the life style of another leucocyte. When these molecules are made by lymphocytes, they can be referred to as *lymphokines*. A more general term for such a molecule, made by one cell that affects another cell, is *cytokine*. Sometimes the type of cell making the cytokine and the cell responding to it belong to separate cell lineages and are thus distinct cell types, in which case the interleukin, lymphokine, or cytokine is said to act in a *paracrine* fashion. In contrast, sometimes a cell makes a molecule and the responding cell is of the same type, in which case the interleukin, lymphokine, or cytokine is said to act in an *autocrine* fashion. Some interleukins act in both an autocrine and paracrine fashion. There are upwards of a hundred such molecules that have now been characterized. We shall later consider critical aspects of how these molecules function. However, the discovery of one of them changed the field in two ways. Firstly, understanding how this molecule functioned provided a paradigm of how one cell type could influence the behavior and fate of another cell type. Secondly, it led to dramatic technical advances as outlined in the next section.

We have seen that lymphocytes, harvested from an animal immunized with OVA, will proliferate when incubated with OVA in vitro. We have also seen that this proliferation of primed T cells requires not only antigen but also APCs. This type of observation was first interpreted, as previously indicated, as being due to the proliferation of the increased number of OVA-specific T cells among the lymphocytes from primed compared to those present in normal animals; it was originally envisaged that this greater prevalence allowed the proliferation of OVA-specific T cells to be detected.

Further observations, however, led to a questioning of the adequacy of this interpretation. It was found that the supernatant of the stimulated T cells, when harvested and added to *unprimed* lymphocytes, resulted in *their* proliferation in the absence of the priming antigen. It became clear with time that stimulation of OVA-specific CD4 T cells with OVA, in the presence of MHC-compatible APC, could result in the production of a molecule called *interleukin-2*, or *IL-2*, that acts as a growth factor for partially activated T cells. Thus, not all the T cells stimulated to proliferate were OVA-specific, even though the production of IL-2 was dependent on both OVA, APC and OVA-specific CD4 T cells. As we shall now see, the availability of a growth factor that can universally support the growth of partially activated T cells was critical for many advances in the field.

The cloning of T cells and the generation and use of B cell and T cell hybridomas

The characterization of IL-2 and the cloning and expression of the IL-2 gene led to the ready availability of recombinant IL-2. This ready availability allowed individuals to grow millions of genetically identical cells from a single, activated CD4 T cell, or from a single, activated CD8 T cell. The single T cell is stimulated with antigen and APC in the presence of IL-2. These populations of T cells, derived from a single activated T cell, are called *T cell clones*. The generation of T cell clones opened up many experimental avenues.

Of somewhat similar use to T cell clones were *B cell* and *T cell hybridomas*. B cell hybridomas were obtained by fusing two diploid cells together to make a hybrid, tetraploid cell, called a *hybridoma*. One cell was an antibody-producing cell, and the other was from a tumor of an antibody-producing cell. Cells of a hybridoma retain the ability to make the particular antibody of the parental antibody-producing cell, and inherit the immortality of the tumor cell parent, resulting in a hybrid that grows continuously and produces copious amounts of homogenous antibody, called a *monoclonal antibody*. Such antibodies, of defined specificity and available in virtually unlimited supply, have become important for research and for clinical purposes.[89] It was for developing this technology that Kohler and Milstein received their Nobel Prize. T cell hybridomas can be made in a similar fashion to B cell hybridomas through fusion of a specific T cell with a cell from a T cell tumor. The T cell hybridoma inherits an antigen receptor derived from its non-cancerous parent and its immortality from its other parent.

The isolation and the structure of the T cell receptor (TcR) and the construction of TcR transgenic mice

The T cell receptor is made in much smaller amounts in a T cell than the amounts of antibody made by an antibody-producing cell. Moreover, much of the analysis of antibody structure involved the use of homogenous myeloma proteins made by B cell tumors. The isolation of the T cell receptor presented formidable problems. However, three different groups found a means of isolating the T cell receptor or their genes. All successful isolations employed T cell tumors or T cell clones or hybridomas. These studies showed that the T cell receptor consists of two chains, called α and the β. Each chain contains two Ig-like domains of about 120 amino acids, a v_α and c_α domain for the α chain, and a v_β and c_β domain for the β chain.

It was found that the diversity of α and β chain genes is generated in a manner similar to that of light and heavy immunoglobulin chain genes, involving DNA joining events of diverse V, D, and J elements, though the elements employed for generating T cell receptor genes are distinct from those involved in generating Ig genes. Successful rearrangement of these elements, to result in functional α and β chains, leads to suppression of further attempts at rearrangement. Thus, the genetics and biology of the T cell receptor have many features in common with those of B cell receptors, i.e. antibodies. It has also been possible to make mice transgenic for rearranged *T cell receptor (TcR)* α and β chains. Such mice are referred to as *TcR transgenic mice*. The presence of the rearranged genes, coding for functional α and β TcR chains, suppresses endogenous rearrangements of α and β DNA elements, so all the T cells of such mice express the same transgenic T cell receptor.

The developmental generation of T cells

On the size of the combining sites of antibodies and of T cell receptors and their biological significance

The size of a combining site, or area of contact between a receptor and its ligand, is a critical parameter for several reasons, and particularly for the way antibodies and T cell receptors bind antigen or peptide/MHC complexes. We can appreciate this by examining various situations.

Consider the imaginary case in which an antibody molecule has a binding site of the size able to recognize one amino acid. There are twenty amino acids commonly present in proteins. A minimum of twenty antibodies would therefore be needed to achieve universality, so that the population of antibodies could react with all conceivable proteins, and so allowing the immune system to achieve universality. In practice, there would be a need for more than twenty antibodies as each amino acid residue could be in a few conformations. However, virtually all such antibodies would be able to bind some self protein or another and, once tolerance to self antigens is established, there would be few if any precursor cells left with antibody receptors able to recognize foreign antigens. This theoretical possibility therefore does not provide a potential basis for how antibodies recognize antigen.

Consider now the second imaginary situation in which we have a larger binding site, recognizing four amino acids, such that a substantial number of antibody molecules remain after those binding self antigens are eliminated. There are twenty amino acids, and so $20 \times 20 \times 20 \times 20 \sim 10^5$ possible combinations of four amino acids, but any set of four amino acids could themselves be arranged in many configurations, say a thousand at a wild but conservative guess. The attribute of universality of the immune system means that all, or at least most of, these combinations can be recognized. Therefore about 10^8 antibodies will be needed to recognize these different *determinants* or *epitopes*, whether foreign or self. Suppose half of these antibodies interact with self, so their corresponding precursor cells are eliminated, leaving 5×10^7 able to react to foreign antigens. The frequency of a specific lymphocyte, able to recognize a particular epitope, would be $1/(5 \times 10^7)$. These figures are very rough, but this does not matter much in light of the considerations arising from the third imaginary situation.

Suppose the area of binding between antigen and antibody is complementary to eight amino acids. How many antibodies would be needed to recognize all possible determinants or epitopes double the previous size? We have estimated there are roughly 10^8 antibodies required to recognize determinants composed of four amino acids. These different areas can each be combined with each other to form distinct areas of eight amino acids in, let us suppose most conservatively, a hundred different ways. Thus, we would need $10^8 \times 10^8 \times 100$ antibody molecules to recognize all determinants of this size, which is 10^{18}. These considerations show how the number of antibodies needed to ensure universality depends critically on the size of the combining site. If too small, the binding site cannot distinguish self from foreign determinants, as illustrated by the first imaginary example. If too large, the number of antibody receptors needed in the repertoire to achieve universality is too large. The implication of the third imaginary example, in the context of Clonal Selection Theory, with a repertoire of 10^{18}, is that there must be at least 10^{18} B cells. However, there are only about 10^{10} cells in an individual mouse. The second imaginary situation is surely closest to the truth. However, one implicit assumption on which these considerations is based is incorrect for B cells. The assumption that the size of the binding site does not vary much between different antibodies is known to be incorrect. Thus, we can raise antibodies specific for artificial haptens about the size of an amino acid, and some antibodies have much greater areas of contact with their antigen than four amino acids. This diversity in size of binding sites makes even approximate, *ab initio* considerations as to the size of the antibody repertoire problematic.

As we have seen, Xray crystallography has revealed how the T cell receptor interacts with the peptide/MHC complex. Such an analysis gives a different picture of the interaction of the TcR with the peptide/MHC complex than that between antibody and antigen. The T cell receptor binds to several amino acid residues of the antigenic peptide as well as to the lips of the MHC groove. The T cell receptor, embedded in the membrane of the T cell, does not interact by itself with the peptide MHC complex, but there are other molecules that bind together in the interaction of the T cell receptor with a peptide/MHC complex of another cell. For example, the CD4 and the CD8 molecules are known to respectively bind to invariant parts of class II and class I MHC molecules. Thus, the T cell receptor on a CD4 T cell binds to a peptide/MHC complex on the surface of an APC facilitated by CD4 molecules binding to the class II MHC molecules. The T cell receptor and co-receptors *dock* onto the peptide/MHC complex, and a rather large and relatively well-defined area of the T cell receptor appears to be in contact with the peptide/MHC complex. The difference in the nature of the two interactions of antibody with antigen, and of T cell receptor with peptide/MHC complexes, means the analysis outlined above, on the relationship between the size of the antibody binding site and the size of the antibody repertoire required to achieve universality, cannot be directly applied to the interaction of T cell receptors and peptide/MHC complexes. However, our realization that the size of the area of contact between the receptor and its ligand dramatically impinges upon repertoire size, if universality is to be attained, is significant in the case of T cells. The consistently large area of contact between T cell receptors and peptide/MHC complexes means that the repertoire is intrinsically very large. The larger the repertoire, the scarcer are the T cells of a given specificity. We shall see in the next section that nature has found a way of partially getting over this problem of extreme scarcity.

Positive and negative selection in the thymus

One of three ways to death for developing T cells: death due to non-functionality

The thymus is the **primary lymphoid organ** in which T cells are generated. There are several distinct steps a pre-T cell must undergo to result in a functional T cell. It first tries to assemble genes coding for functional TcR α and β chains. Many attempts at creating functional TcR α and β chains by DNA joining events of TcR genetic elements fail to give rise to functional TcR α and β chains. Unsuccessful attempts at generating functional α and β TcR chains is the earliest way to death for a pre-T cell undergoing the processes associated with T cell development.

One of three ways to death for developing T cells: death for self protection

Another pathway to death was anticipated in terms of the ideas incorporated in Lederberg's model of lymphocyte generation. The T cells that acquire a functional receptor able to recognize a self peptide/MHC complex of a nominal self antigen prevalent in the thymus are known to be killed. This purging of T cells with anti-self activity is called **negative selection**. It is of major importance in achieving tolerance.[90] This negative selection of T cells that occurs in the thymus is called **central tolerance.** However, not all self antigens and their corresponding peptide/MHC complexes are present in the thymus at sufficient levels to purge all the T cells specific for the nominal self antigen. For example, there will be T cells generated that are specific for nominal antigens predominantly expressed in highly specialized cells present in organs. An example is insulin made by the β-islet cells of the pancreas. Insulin is usually a target of attack in autoimmune diabetes. Normally, some T cells specific for the nominal antigen insulin escape the process of central tolerance,[90] and these T cells emigrate to the **periphery** where they circulate around the body or lodge in **secondary lymphoid organs** where immune responses are mounted, such as lymph nodes and spleen. These functional T cells that emigrate from the thymus are called **peripheral T cells**. These insulin-specific, peripheral T cells are thought to be inactivated by insulin in healthy individuals in a manner we shall speculate upon later. It is an unusual circumstance that leads to their activation and, when this occurs to a sufficient degree, it can lead to frank autoimmunity. The successful inactivation of such peripheral T cells leads to a state of **peripheral tolerance**. Self antigens, not present in the thymus at the level required to cause complete purging of their specific T cells by the process of central tolerance, are called **peripheral self antigens**.

One of three ways to death for developing T cells: death by neglect

A third way to death for developing T cells became apparent as people tried to understand the steps involved in the generation of T cells. It is helpful to preface these studies by considering some implications of some known and remarkable facts.

The greatest variablilty between the different class I MHC molecules present within a species, such as mice, occurs in the floor, walls, and lips of the grooves of these molecules. As we have noted, different peptides bind to the grooves of different MHC molecules. We have also seen that the TcR binds to both a peptide, itself bound to the groove of the MHC molecule, and the lips of the groove. As the lips of the grooves of different class I MHC molecules are different from one another, a TcR that fits well the lips of one class I MHC molecule is unlikely to bind well to the lips of a different class I MHC molecule. In fact, many of the newly generated cells in the thymus, that have functional T cell receptors, have receptors that cannot bind to the lips of any of the class I MHC molecules present on host cells. These cells therefore have no potential to be useful; they are *useless*. Other newly generated cells, that have functional T cell receptors, have receptors that can bind to the lips of the grooves of one or another class I MHC molecule present on host cells. These cells have the potential of being *useful*. The studies to be outlined delineate a process by which the useless T cells are allowed to die, and are in effect discarded, and the useful T cells are given a kiss of life to survive, and emigrate to the periphery.

Recall that thymectomy of mice at birth results in adult mice with a deficiency of Thy1+ T cells. Thymectomy at three weeks of age has little effect because, by this time, many T cells have been generated in, and emigrated from, the thymus to become the pool of T cells that travels around the body, residing in the spleen and lymph nodes. These peripheral T cells multiply and die at a measurable rate. If the mouse is heavily irradiated, the DNA of the lymphocytes is damaged, and when they attempt to divide, the lymphocytes die. Most of the white blood cells of a lethally irradiated mouse rapidly disappear after radiation and the mouse will die if not given unirradiated, syngeneic **stem cells**, which allow the diverse array of white blood cells to be regenerated. Stem cells are present in and are typically obtained from the bone marrow.

When thymectomised mice are heavily irradiated and reconstituted with syngeneic bone marrow stem cells, all the diverse array of different white blood cells are generated except for Thy1 bearing cells. The thymic environment is needed for bone marrow stem cells to give rise to Thy1+ T cells. This can be directly shown. A thymus, heavily irradiated to ensure the future death of its Thy1+ cells, is placed under the kidney capsule of a syngeneic, thymectomised mouse. The irradiation of the thymus gland will result in the rapid death of the Thy1+ cells, but will leave intact for weeks many non-dividing epithelial cells of the thymus. When this thymectomised mouse, given an irradiated thymus gland, is itself irradiated and reconstituted with syngeneic bone marrow stem cells, Thy1+ T cells are generated. The grafted thymic epithelial cells under the kidney capsule provide the environment allowing bone marrow stem cells to give rise to Thy1+ T cells.

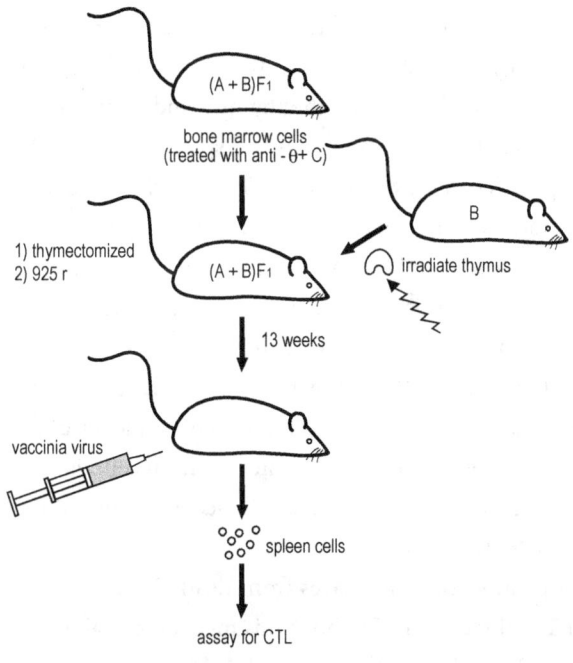

Fig. 28. The type of experiment that demonstrates the *positive selection* of thymocytes. CTL are generated against vaccinia virus infected B but not against vaccinia virus infected A targets.

A variation of this experiment is outlined in Figure 28. The experiment involves the use of mice bred from two different inbred strains, A and B, represented genetically as A/A and B/B, and with different MHC haplotypes. Thus, at a given genetic locus, we can in general represent the allele of the A strain as A, and of the B strain as B, and so the alleles of the so-called F1 cross, obtained on mating A and B strain mice together, as A/B. These A/B mice were thymectomised, given irradiated thymic tissue from strain B mice, after which they were lethally irradiated and reconstituted with A/B bone marrow stem cells. After six weeks, during which time the stem cells give rise to T cells, the mice were infected by vaccinia virus, and the nature of their CTL examined, see Figure 28. Normally, infection of A/B mice results in CTL that lyse virally-infected A and virally-infected B targets, but not uninfected A and B targets. However, in the A/B mice that had B-type thymus tissue, potent CTL are generated upon infection with vaccinia virus that lyse vaccinia virus infected B targets, but not virally infected A targets.[91] Why not? The only significant difference between a normal A/B mouse and the A/B mouse constructed to have only B-type thymic epithelial cells is that the A/B mouse has A/B thymic epithelial cells.

Such observations led to the proposal that pre-T cells are generated in A/B mice with diverse TcRs, some of which can bind A and others of which can bind B type MHC molecules, and yet others that can only bind to other class I MHC molecules that are not present in the A/B mouse. All developing T cells are envisaged to die from neglect if not given the kiss of life when their TcRs bind to MHC molecules present on certain thymic epithelial cells. They can then, if given this kiss of life, differentiate into a mature T cell, exit the thymus, and emigrate into the **secondary lymphoid organs**, such as the lymph nodes and spleen. In the A/B mouse with only B type thymic epithelial cells, the same pre-T cells will be generated with diverse TcRs, as are generated in a normal A/B mouse, but it appears that only those pre-T cells with TcRs able to recognize B MHC molecules on thymic epithelial cells are given the kiss of life, differentiate into mature T cells, and emigrate to the periphery. Thus, in these latter mice, there will not be mature T cells with TcRs that can recognize vaccinia viral peptides in the context of A-type MHC molecules. This process, in which thymocytes are given the kiss of life as a consequence of their TcRs binding to MHC molecules present on certain thymic epithelial cells, is called **positive selection**. The death of cells, with functional T cell receptors but not positively selected, is referred to as **death by neglect.**

Suppose that positive selection saves from death 5% of developing T cells that have functional T cell receptors. Without this mechanism, all the T cells generated with functional TcRs, including those with the ability to recognize peptides in the context of non-self MHC molecules, would have to be saved to keep those recognizing peptides in the context of self MHC molecules that are needed for defense. In this case, the frequency of T cells specific for a given peptide in the context of host MHC molecules would be about twenty times lower than it is with positive selection. In summary, positive selection allows the repertoire in the periphery to be reduced by about twenty-fold, getting rid of those T cells with a TcR that do not recognize self MHC molecules, and so are useless in providing protection against invaders. A further feature of positive selection is interesting and important. The pre-T cells bear both CD4 and CD8 molecules on their surface. When their TcR binds to class I MHC molecules during the process of positive selection, the pre-cell gives rise to mature T cells bearing only CD8, and when the TcR of the pre-T cell binds to class II MHC molecules, the pre-T cell gives rise to mature CD4 T cells. This is why there is a correlation of the CD4/CD8 phenotype of mature T cells with the restriction specificity of their TcR.

How can the receptors of peripheral T cells interact with cells presenting peptide/MHC complexes?

There are typically of the order of 100,000 class I MHC molecules on the surface of a cell, and these might present at least a thousand different peptides. Consider how a CTL interacts with its targets. Its receptors have to find some of those few peptide/MHC complexes that they can bind to and form a stable *immunological synapse* with the target cell, a series of events needed before the CTL can deliver its lethal blow. Another miracle!

It has been found that peripheral CD4 T cells and CD8 T cells need to respectively interact with class II and class I MHC molecules in order to survive over days, even in the absence of the nominal antigen for which they are specific. For example, a population of CD4 T cells transferred into a class II MHC KO mouse, which has no class II MHC molecules, is not sustained, whereas it is on transfer to a normal mouse. Such observations have led to the recognition that the survival of peripheral T cells requires some interaction of their receptors with MHC molecules, a process referred to as *tonic signaling*. Tonic signaling could be somewhat analogous to positive selection in the thymus. The T cell receptors would most often have to interact with MHC molecules bearing less bulky peptides than the *target peptide* their receptors recognize as a "peptide/MHC complex", for steric reasons. The existence of tonic signaling suggests that the receptors of surviving T cells interact almost well enough, but not quite well enough, with bystander MHC molecules to form a stable immunological synapse. This presumably explains how a few receptors of the T cell, interacting with the appropriate target peptide/MHC complexes, can trigger stable synapse formation, as the T cells are continually on the brink of being able to form a synapse without their receptors binding to target peptide/MHC complexes. This mechanism is consistent with the finding that T cell receptors within synapses interact with MHC molecules not presenting the target peptide.

How are anti-self antibody precursor cells ablated?

The insights into how the GOD of antibody genes is achieved again led to the acknowledgment that some genes will code for antibodies directed at self antigens. How is such anti-self reactivity controlled? In mature humans, B cells are generated in the bone marrow and anti-self B cells eliminated by a Lederberg-type mechanism, a process naturally referred to as central tolerance. Thus, there is a need for a mechanism of peripheral tolerance for those peripheral self antigens not present in the bone marrow at a level to ensure central tolerance. We have already talked about how the inactivation of antibody precursor cells specific for peripheral self antigens might be achieved: the peripheral self antigen will inactivate the pertinent B cells if there are no, or an insufficiency of, specific T helper cells. We should only add here that functional Ig genes can also rapidly mutate, so that B cells, specific for and stimulated by a foreign antigen, can give rise to B cells with anti-self activity. This fact further highlights the need for continuous surveillance and silencing of mature anti-self B cells throughout life. We shall shortly describe evidence strongly supporting the one cell/multiple cell model for the antigen-dependent inactivation/activation of B cells. This model results in the continuous purging of anti-self B cells as they arise.

Cellular Interactions

The nature of the B cell/T helper cell interaction

We have seen that primed T helper cells can, when stimulated with the nominal antigen in the presence of APC, produce various molecules including IL-2, which can affect the lifestyle of other lymphocytes. Some of these factors, produced by Th cells, cause B cells to divide, and others cause the progeny of activated B cells to differentiate into antibody secreting cells. Such observations led to the realization that there was more to the role of T helper cells than "presenting antigen", as envisaged in the original antigen-bridge model of the interaction between B cells and T helper cells. These interleukins, lymphokines, or cytokines are not antigen-specific, and act upon target cells in a responsive mode, independently of the specificity of the target cell.

The B cells specific for a particular antigen are scarce, and so their activation is difficult to study. Immunologists overcame this scarcity problem in various ways. One, already outlined, was to employ B cells from mice transgenic for Ig genes. Another earlier approach was to find an analogue of antigen, in its role of interacting with the B cell *surface Ig (sIg)* receptor, which would act on all B cells. They found that certain antibodies to mouse Ig, i.e. to the antibody receptor on murine B cells, caused the B cells to divide. It was clear that the binding of such antibodies to the sIg receptor generated a signal, probably an approximate analogue of signal 1. The possibility existed that this sIg-mediated signal, together with short-range lymphokines, produced on stimulation of activated Th cells with antigen, would activate B cells as efficiently as Th cells, if signal 2 was indeed solely mediated by such lymphokines. However, although such soluble-mediators increased the production of antibody-producing cells under some circumstances, they did so less efficiently than activated Th cells. This led some to suppose that CD4 T cells helped the activation of B cells by doing more than the production and delivery of cytokines.

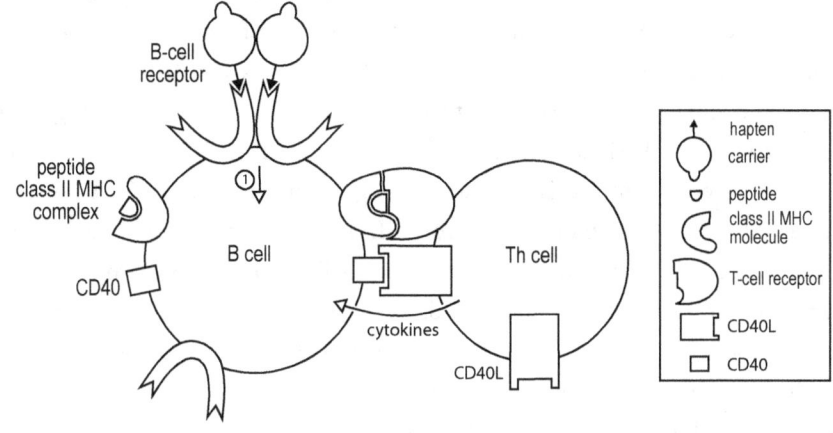

Fig. 29. Model for the activation of B cells, emphasizing the molecular nature of signal 2

It was found that the addition of a preparation of membrane fragments produced from activated Th cells could further enhance the activation of B cells to generate antibody-producing cells. Membrane preparations from *activated* CD4 T cells were much more effective than membrane preparations from *unactivated* CD4 T cells. Such observations led to the tentative proposal that signal 2 is mediated in part by lymphokines produced by activated Th cells and in part by an interaction between molecules expressed on the membranes of activated Th cells and counter receptors present on B cell membranes. A further series of observations supported this insight and further defined the nature of the membrane/membrane interaction. The molecule, referred to as **CD40 ligand (CD40L)**, is present on activated but not on unactivated CD4 T cells. The CD40L binds to **CD40**, found on both unactivated and activated B cells, as well as some others cells, predominantly APC. The activity of the membrane fragments from activated Th cells, in aiding the activation of B cells in conjunction with anti-mouse Ig and lymphokines produced by activated Th cells, could be replaced by some anti-CD40 antibodies. These observations led to the tentative suggestion that the membrane/membrane interaction between activated Th cells and B cells, and required for optimal B cell activation, merely involves an interaction between CD40L and CD40,[92] see Figure 29. Most interestingly, some people who are genetically deficient in producing antibody have mutations in the gene coding for CD40, and others in the gene encoding CD40L, and yet others have other deficits. Parallel observations have been made in mice deficient in either CD40 or CD40L. These further observations support the model outlined in Figure 29.

The activation/inactivation of B cells and of CD8 T cells

A number of studies have shown that B cells are inactivated when they interact with antigen in the absence of CD4 T cells.[93,94] Some of the most striking studies employ B cells obtained from mice transgenic for Ig chains that code for an antibody of known specificity. For example, Chris Goodnow and colleagues constructed mice with transgenes coding for IgM molecules that are specific for the antigen *hen egg lysozyme (HEL)*. They also made other mice transgenic for HEL. These HEL transgenic mice would therefore regard HEL as a self antigen. They found that IgM anti-HEL transgenic B cells, when transferred into HEL transgenic mice, could be activated to produce anti-HEL antibody, if immediately challenged with HEL coupled to an immunogenic carrier. The response was also examined after a similar challenge three days after transfer of the IgM anti-HEL B cells into HEL transgenic or into normal mice. Immunization of normal mice with a HEL-carrier conjugate, which had been given IgM anti-HEL transgenic B cells three days earlier, led to the production of substantial anti-HEL antibody, whereas the similar transfer of the IgM anti-HEL B cells into HEL-transgenic mice, with a similar antigen challenge three days later, did not result in the production of anti-HEL antibody. The B cells had been inactivated, silenced, during their three-day residence in HEL-transgenic mice.[94] This inactivation was presumably because these HEL-specific B cells interacted with HEL in the absence of HEL-specific helper T cells.

Other studies show that the optimal activation of CD8 T cells to become CTL requires in most systems the presence of activated CD4 T cells, acting via the operational recognition of linked epitopes.[95] In the absence of such T cell help, the antigen inactivates the CD8 T cells.[96] Thus the presence/absence of activated CD4 T cells specific for a nominal antigen has a profound influence on whether antigen activates/inactivates B cells and CD8 T cells. The general nature of these observations was anticipated by our original Two Signal Model of lymphocyte activation. They show that CD4 T cells are the guardians over the fate of most other lymphocytes when these lymphocytes encounter antigen. This recognition of their central role has led in recent times to the prominence of the question, what controls whether antigen activates or inactivates CD4 T cells?

What determines whether antigen activates or inactivates CD4 T cells?

This is for me one of the prime questions of the field and is at the fore of current debate. It is pivotal not only in itself but because potential answers set the scene for considering other critical questions. We need to examine this question from diverse perspectives.

The original Two Signal Model

I believe our Two Signal Model of lymphocyte activation[47] was the first such model for the activation of CD4 T cells. All contemporary models for the activation of CD4 T cells bear some resemblance to the original. There is a consensus over the nature of signal 1, generated when the *T cell receptor (TcR)* interacts with a peptide/class II MHC complex on an antigen presenting cell (APC). Most agree that this signal, when generated by itself, leads to the elimination or silencing of the pTh cell. The differences between the models rest on what is envisaged to be necessary for the generation and delivery of the critical signal 2, additionally required for T cell activation. There are subtle, mechanistic differences as to what these requirements are in the different models. These proposals have to my mind profoundly different physiological consequences. I think the best way of appreciating the issues involved is to follow historically the considerations that led to the different models.

I start with our original two signal proposal, made in 1970.[47] In modern parlance, we proposed that Th cells are required to help in the activation of *precursor Th (pTh) cells*. Our proposal is depicted in its essential form in Figure 16. This was a time when the MHC-restricted nature of the recognition of antigen by T cells was not a possibility anyone had thought of, and so the model is expressed within the context that the T cell receptor recognizes intact antigen. Our Two Signal Model of lymphocyte activation appeared to have rather little impact, as assessed by citations, until its pertinence to B cell activation and inactivation was recognized.[94] Our paper is often quoted as though we only address the requirements to activate and inactivate B cells, even though we explicitly proposed that the activation of a pTh cell required its antigen-mediated interaction with other Th cells, mediated by a two signal mechanism.[47] In addition, we and others provided over the years evidence that the activation of CD4 T cells requires, or is facilitated by, CD4 T cell collaboration.[97-99]

The Constitutive Model

Lafferty and Cunningham were trying to understand in the early 1970s how MHC-specific CD4 and CD8 T cells are activated by stimulatory cells that bear foreign MHC molecules. They noted a number of circumstances where the ability of such cells, to stimulate the generation of T cells specific for these MHC molecules, did not correlate with the level of expression of the foreign MHC molecules on the stimulating cells. The expression of foreign MHC molecules was necessary but insufficient to activate T cells. Lafferty and Cunningham referred to cells bearing foreign MHC molecules and able to stimulate vigorous allogeneic responses as having an **S⁺ phenotype**, whereas those foreign MHC-bearing cells, unable to efficiently generate anti-MHC T cells, were designated as having an S⁻ phenotype,[100] see Figure 30. They found, for example, that gamma irradiation of an S⁺ cell could destroy its S⁺ phenotype without affecting its expression of MHC molecules. Lafferty and colleagues took their analysis further, and showed that the rapid rejection of MHC-incompatible grafts is due to the presence, within the graft, of cells with an S⁺ phenotype. They found that some grafts, depleted of S⁺ cells by culturing the grafts under conditions inimical to the S⁺ cells and resulting in their death, but not so inimical to the bulk of the cells constituting the foreign graft, rendered the graft, on being transplanted onto an MHC-incompatible host, dramatically less able to induce an immune response and so be rejected. They drew the inference that a foreign MHC-bearing cell, present in the uncultured graft and able to stimulate the generation of anti-MHC T cells, did "something more" than than simply present foreign MHC allo-antigens. Indeed, there was considerable hope at one time that this procedure of culturing grafts might constitute a way of overcoming the transplantation problem.

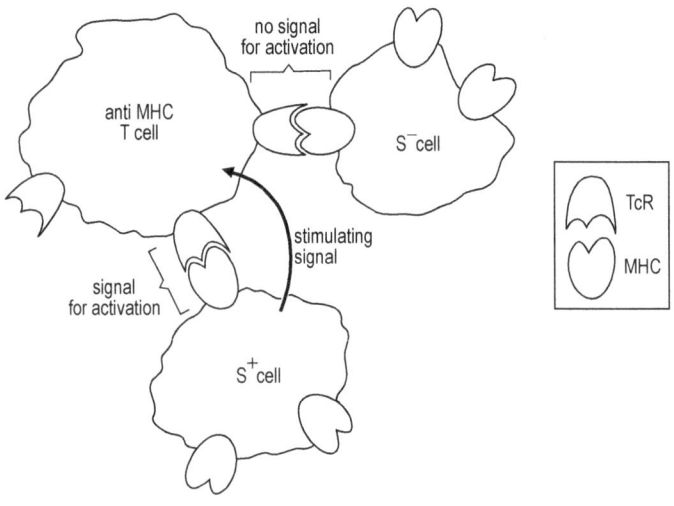

Fig. 30. The Lafferty/Cunningham Model for the Activation of T cells

The model proposed by Lafferty and Cunningham was supported, and given a more defined form, by observations made some years later. Immunologists wanted to discover how specific CD4 T cells are activated and inactivated by antigen.

We have seen that the scarcity of unprimed CD4 T cells, specific for a particular antigen in a normal animal, poses a formidable technical problem for determining the requirements to activate/inactivate them. Quill and Schwartz took rested cells from Th1 clones and examined the conditions under which these T cells could be activated and inactivated by antigen.[101] They knew that these T cells recognized a particular peptide p bound to a particular class II MHC restriction element, R, i.e. their TcR recognized the p/R complex. They thought studies on the requirements to activate these rested T cells might provide clues as to how naïve CD4 T cells are activated. These rested CD4 T cells could be stimulated by p, in the presence of APC bearing R, to multiply and generate progeny that produced Th1 lymphokines. These investigators chemically isolated R from APC, and inserted R into an artificial membrane, consisting of a lipid bilayer devoid of other proteins. They then examined whether these rested Th1 cells could be activated when exposed to p bound to the R molecules that were inserted into this lipid bilayer. Such exposure resulted in neither the division of the T cells nor in the production of lymphokines. It seemed that more than an interaction of the TcR with the p/R complex was required to activate these CD4 T cells, a deduction that paralleled the conclusion reached by Lafferty and Cunningham. However, they did find that the CD4 T cells had altered properties if they interacted with the p/R complex over a period of two days. These exposed CD4 T cells could no longer be stimulated by APC, loaded with p, to produce progeny that secreted Th1 lymphokines. Quill and Schwartz suggested on these grounds that signal 1, when generated alone, could inactivate these CD4 T cells.[101]

This proposal was supported by further observations. They led to the suggestion that the interaction of B7 molecules, expressed by APC, with CD28 molecules expressed by the CD4 T cell, generaterd a signal required for CD4 T cell activation. It was possible to add molecules that bind to the B7 molecules on the APC, and so block their interaction with CD28 molecules on the T cell, in which case the CD4 T cells were not activated, but inactivated. Moreover, the additional presence of certain antibody molecules to CD28 results in the activation of these CD4 T cells. It thus appeared that this anti-CD28 antibody mimicked the effects on the T cell of the interaction of the CD4 T cell's CD28 molecules with B7 APC molecules. Activation also required the presence of the peptide "p". It thus seemed that the activation of these CD4 T cells requires signal 1 and a second signal, signal 2 or the ***costimulatory (CoS) signal***, mediated by the interaction of the CD28 molecule on the CD4 T cell with B7 molecules on the APC, see Figure 31.

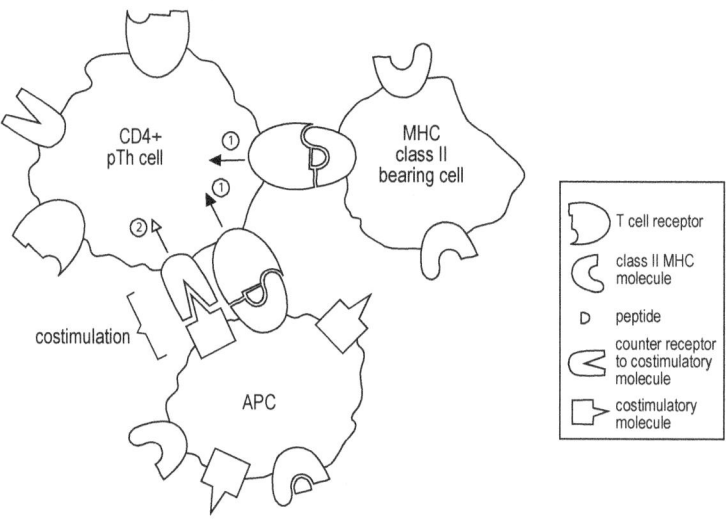

Fig. 31. The Constitutive Model of CD4 T cell activation

The B7 molecules on the APC are referred to as costimulatory molecules, and the CD28 molecules as their counter receptors. The CD28 molecule is present on naïve T cells as are the B7 molecules on some mature APC. A number of interactions between distinct molecules on APCs and their counter receptors on T cells can lead to a signal that cooperates with signal 1 in the activation of CD4 T cells. Such membrane/membrane interactions, leading to second signals, are collectively referred to as *costimulatory signals*, and the interacting molecules on the APC and the CD4 T cell are respectively referred to as the costimulatory molecule and its counter receptor. A characteristic of B7 molecules is their substantial presence on most mature APC, such as macrophages. The expression of other costimulatory molecules is clearly regulated, requiring the cells to be activated in some way that we shall shortly discuss. The expression of B7 molecules on macrophages is not heavily regulated and is constitutively expressed. The envisaged critical role of these constitutively expressed CoS molecules in this model is why I refer to this model as the *Constitutive Model of CD4 T cell activation*, see Figure 31.

This Constitutive Model of CD4 T cell activation was critical in the development of ideas as to how CD4 T cells are activated. However, it is now no longer seriously entertained. This is in large part because mature macrophages express B7 molecules constitutively and can present peptides derived from both peripheral self antigens and from foreign antigens. Where is there scope in the context of this model for understanding how peripheral self antigens can inactivate their corresponding CD4 T cells and foreign antigens can activate theirs?

The Infectious/Danger Models

We are now in a position to describe and discuss how the most widely accepted model of CD4 T cell activation arose. Charlie Janeway was impressed by the number of observations that are most naturally explained in terms of a requirement for two signals to activate lymphocytes, including CD4 T cells.[102] However, he pointed out that it is often difficult to raise immunity in mice against a highly purified protein, such as BSA, and that immunologists most often resort to the use of microbial adjuvants in order to raise immunity against such antigens. First among these adjuvants is *Freund's Complete Adjuvant (CFA)* that contains dead mycobacteria. Janeway proposed that such adjuvants are required to generate immune responses against such antigens, and he famously dubbed this lack of acknowledgment as "the immunologists' dirty little secret."[102] There is no doubt that dead mycobacteria contain a number of *pathogen-associated molecular patterns (PAMPs)*. Janeway proposed that the interaction of such PAMPs with *pattern recognition receptors (PRR)* was necessary for the full and appropriate expression of the APC's costimulatory molecules that are required for the activation of naïve CD4 T cells. He further suggested that exposure to antigen in the absence of PAMPs inactivated the corresponding CD4 T cells. He made the radical suggestion that the immune system really does not discriminate between self and non-self at the level of peripheral CD4 T cells; he proposed it would be much more accurate to state that the immune system discriminates "non-infectious" entities from PAMP-expressing "infectious" entities, see Figure 32. The striking implication of this view is that non-infectious but foreign antigens would not be immunogenic but tolerogenic.

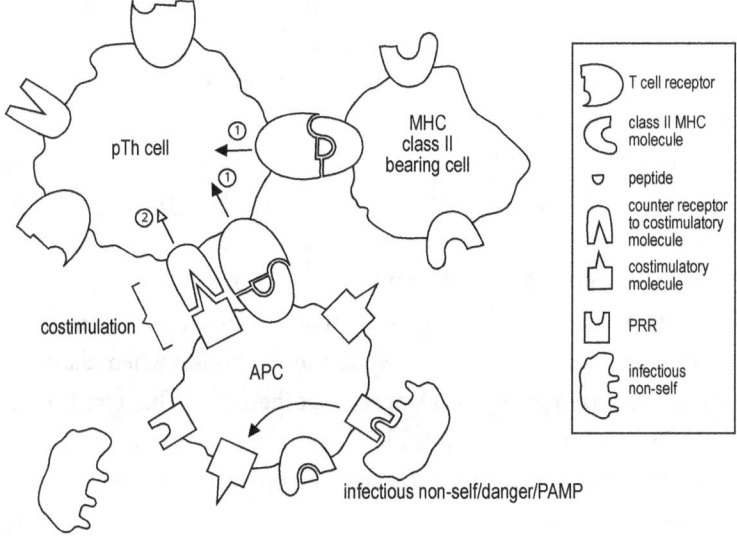

Fig. 32. The infectious/danger model of CD4 T cell activation.

To fully appreciate Janeway's proposal, for both its virtues and deficiencies, it is helpful to better understand the function of some APC at the earliest stages of an immune response.

There are many immature APC in various parts of our body, and particularly in the skin, potential sites of entry by some pathogens when skin abrasions occur. These potential APC are for the most part *immature dendritic cells (DC)* that readily injest, by various means, matter with which they come into contact. *Immature DC* can develop into *mature DC.* This maturation process results in three changes in the nature of the DC: loss of phagocytic function, greater surface expression of class II MHC molecules as well as B7 costimulatory molecules and the ability to migrate to the closest draining lymph node, where cells of the immune system interact to generate immune responses. This process of DC maturation and migration appears to provide an essential surveillance function. These events collectively bring information into the draining lymph node about the spectrum of antigens present at the peripheral site where the DC matured and, if appropriate, lead to the initiation of an immune response against any foreign antigen present at this peripheral site. Janeway's view is that the full maturation of APC, so that they can activate CD4 T cells, requires them to receive a signal following the interaction of a PAMP with a PRR, most often but not necessarily present on the surface or inside the APC.

There is no doubt that various PAMPs can at least facilitate and accelerate the maturation of DC, but whether the maturation process can take place as a slower pace in the absence of PAMPs seems to me to be an open question. Whatever the case, Janeway envisages that PAMPs are necessary to activate CD4 T cells. I consider Janeway's proposal that the immune system discriminates, at the level of CD4 T cells, between noninfectious and infectious entities, rather than between self and nonself, as a radical change, and so should be carefully scrutinized.

A weakness of Janeway's model is that it appears to be possible to generate immune responses to some foreign, sterile vertebrate antigens without adjuvant. Such antigens are not expected to bear PAMPs. An example in mice would be sterile sheep red blood cells. In addition, it is unclear how Janeway's proposal can be reconciled with the rejection of allogeneic transplants. Polly Matzinger extended Janeway's proposal. She suggested that the body has a more general means of detecting when an immune response is beneficial than the means envisaged by Janeway, and proposed that the body could initiate immune responses when "danger" occurs. This proposal is referred to as the "Danger Hypothesis."[103] What constitutes danger is not very clear to me, but it probably includes all the situations envisaged by Janeway, and others where there is stress, such as when our skin is exposed to a sharp object, or when skin is transplanted. I sometimes need to refer jointly to Janeway's and Matzinger's models. I refer to them on these occasions as the *PAMP/Danger*

Model. This shorthand should not be taken to mean that the two models are identical. They are clearly different. However, they do have several common features.

There is no doubt that these models, particularly Janeway's, have stimulated an incredible amount of interest and a great number of revealing studies. It has been shown that PAMPs interact with PRR, leading to such effects as the increased expression of costimulatory molecules on APC. Thus, there is no doubt that interactions between PAMPs and PRRs affect the generation of immune responses in diverse ways. It makes sense that if one, or several, PRR are triggered, indicating the presence of microbes/parasites inside the body, that evolution would have found the means of employing these signals of alarm to good effect by, for example, increasing the rapidity and intensity of immune responses. Although I disagree with the plausibility of the PAMP/Danger Models for the activation of CD4 T cells, in the strict sense in which they were first proposed, I embrace the evidence that innate mechanisms of defense contribute to the regulation of immune responses, not only on an observational basis, but because this makes much physiological sense to me. I shall elaborate on this later, with the hope that I can dispel the appearance of any equivocation.

Epitope spreading and negative regulation of CD4 T cell activation

One feature of the original Two Signal Model of lymphocyte activation, as it pertains to CD4 T cells, is that the CD4 T cell cooperation involved in CD4 T cell activation constitutes a positive feedback loop. In the presence of a constant source of antigen, this loop, as described, would result in an escalating cascade of CD4 T cell interactions and in the further activation of helper CD4 T cells. Indeed, there is evidence for such a cascade. Thus, the original Two Signal Model naturally accounts for a phenomenon called *epitope-spreading*. Consider a protein Q that is processed to generate peptides q_1, q_2, q_3, ... q_n, that bind to host class II MHC molecules and for which there are specific CD4 T cells. It has been found that when, for example, Q is a peripheral self antigen, the deliberate generation of CD4 T cells specific for q_2 can result in time in the spontaneous generation of CD4 T cells specific for other q peptides. It appears that the presence of CD4 T cells specific for one peptide or epitope of an antigen helps in the generation of CD4 T cells specific for other peptides or epitopes of the antigen, hence the name of this phenomenon as *epitope-spreading*. Epitope spreading is observed in both animal[104,105] and human[105] autoimmune disease. In the mouse model of autoimmune diabetes, in the strain known as *non-obese diabetic (NOD)*, autoreactivity spontaneously develops against the β-islet cells of the pancreas that produce insulin. The autoimmunity of some NOD mice shows striking similarities to that seen in human autoimmune diabetes.[5] It has been shown that NOD mice develop CD4 T cell autoimmunity to different β-islet antigens in a pretty strict temporal order.[104] Injection into the thymus of very young mice, of the autoantigen against which autoreactive CD4 T cells first appear, blocks the development of autoreactive CD4 T cells specific for this first antigen as well the CD4 T cells specific for the other antigens. These observations are naturally accounted for if there is a functional role of CD4 T cell cooperation in the spreading of autoimmunity.

The uncontrolled expansion of CD4 T cells, which would follow the unrestrained CD4 T cell facilitation of CD4 T cell activation, seems implausible. Thus the phenomenon of epitope spreading, though supporting the idea of CD4 T cell cooperation, also illustrates its untoward consequences. These are particularly evident when considering the potential expansion of CD4 T cells specific for self antigens, as these antigens are replenished even as they are damaged and removed by an autoimmune response.

It is clear that the activation of CD4 T cells is regulated by various feedback mechanisms. A central mechanism is mediated by the **CTLA-4 antigen**. This antigen is present on all activated CD4 and CD8 T cells but absent from naïve, unstimulated CD4 and CD8 T cells. The CTLA-4 antigen is evolutionarily related to the CD28 molecule, the T cell's counter receptor to the APC's B7 costimuatory molecules. Moreover, the CTLA-4 molecule also binds B7 molecules, but with considerably greater affinity than CD28. Thus when CD4 T cells are activated, and they express CTLA-4, the CTLA-4 molecules block any CD28/B7 interactions that would otherwise be involved in CD4 T cell activation. The importance of CTLA-4 in regulating CD4 T cell activation is apparent from the phenotype of mice constructed to be deficient in CTLA-4 molecules. Young deficient mice have increased levels of activated CD4 T cells, some of which infiltrate organs, and the mice die by about three weeks of age. It is believed that they die from an unrestrained inflammatory response, most probably directed in part against self antigens.[106]

Another protein, expressed on activated T cells and structurally and evolutionarily related to CD28 and CTLA-4, is the **PD1 (programmed death-1) molecule**. Its ligands, PDL1 and PDL2, are broadly expressed on various cells, including APC. The importance of the PD1 molecule is evident from PD-1 deficient mice that have some late-appearing and relatively mild symptoms of autoimmunity. It is likely that the inactivation of partially activated T cells can be facilitated by the interaction of the T cell's PD1 molecules with its ligands,[107] though I personally think that the generation of signal 1 alone may be sufficient to inactivate naïve T cells. Otherwise, I would expect mice unable to express PD1 to suffer from very severe autoimmunity.

An introduction to the existence of distinct Th cell subsets

I explained at the beginning of this book why I believe an understanding of the basis of immune class regulation is likely critical for designing strategies to prevent or to treat various clinical sitations related to allergies, infectious diseases, and cancer. The observations I shall shortly outline are of cardinal importance in two respects. They constitute the first real break-through in understanding at the molecular level how different types of CD4 T cell are associated with, and help to control, the class of immunity induced. Second, although this conceptual framework originally involved the characterization of just two Th subsets, the Th1 and Th2 subsets, associated respectively (and roughly) with cell-mediated and humoral responses, the idea that different Th subsets control the class of immunity has proved to be more generally useful as our understanding of the sophistication of immune class regulation has developed.

It was insightful to employ cells from different T cell clones to examine what functions they have. Such studies gave rise to the realization that different types of Th cells have different functions, which are largely due to the different spectrum of lymphokines these Th cells produce upon stimulation with their nominal antigen. Most interestingly, it was found that many of these Th clones, derived from in vivo immunized lymphocytes that had been propagated by in vitro stimulation with the nominal antigen, tend to fall into two groups, the famous Th1 and Th2 clones.

Coffman and Mosmann showed that cells of *Th1 clones* produce IL-2 and IFN-γ, among other lymphokines, and cells of *Th2 clones* characteristically produce IL-4 and IL-10.[108] The lymphokine IL-2 is employed for cloning T cells because all activated T cells respond to IL-2 as a growth factor, including the type of Th cell that produces it. This lymphokine thus functions in both an *autocrine* as well as a *paracrine* manner. Cells belonging to Th1 clones mediate a DTH reaction upon injection with antigen under the skin of a syngeneic mouse. In addition, CD4 T cells from mice immunized to express DTH against an antigen, produce IL-2 and IFN-γ on stimulation with this nominal antigen. The parallel nature of these two sets of observations, one made with cells from T cell clones, the other with in vivo sensitized Th cells, gave rise to the notion that cells of Th1 clones "represent" those in vivo generated T cells mediating classical DTH responses, and that such cells are committed to produce upon stimulation IL-2 and IFN-γ.

The properties of the lymphokines produced by cells of Th2 clones are also enlightening. The lymphokine IL-4 acts in an autocrine manner, as it is produced by and is a growth factor for Th2 cells. The addition of IL4 to spleen cells results, under certain conditions, in a greater production by the spleen cells of IgG$_1$ and IgE antibody. Somewhat similarly, the addition of IFN-γ to mouse spleen cells can result

in greater IgG$_{2a}$ antibody production. We thus begin to understand how the generation of distinct CD4 T cell subsets leads to and reflects the generation of different
classes of immunity.

In some in vivo immune responses, Th1-like cells are predominant and, in others,
Th2-like cells predominate. Such responses will be referred to as predominant Th1
and predominant Th2 responses. In other responses, the immune lymphocytes
are stimulated by antigen to produce lymphokines characteristic of both Th1 and
of Th2 cells. We refer to such immune responses as having a mixed Th1/Th2 phenotype. In yet other responses, the CD4 T cells generated do not fit into the Th1/
Th2 paradigm.

Ideas on the nature of the Decision Criterion determining the Th1/Th2 phenotype of immune responses

The Threshold Hypothesis has had little impact on the field, despite being put
forward in the early 1970s as a description of the mechanism controlling the Th1/
Th2 phenotype of immune responses. However, observations made in the early
1990s and later were influential in the formulation of current ideas as to the nature
of the decision criterion controlling the Th1/Th2 phenotype of immune responses.
We shall outline here the nature and basis of three ideas that are prominent today,
namely the **Cytokine Milieu Hypothesis**, the **PAMP Hypothesis**, and the **DC
Hypothesis.** We shall later discuss whether these different potential insights are in
conflict or how they might be mutually accommodated.

A seminal paper examined the conditions under which *Listeria monocytogenes*, an
intracellular, human bacterial pathogen that thrives in macrophages, could influence
the Th1/Th2 phenotype of the activated CD4 T cells generated. It was shown that
dead bacteria could stimulate the *in vitro* production by macrophages of IL-12. The
presence of IL-12 could, in turn, favor the generation of Th1 cells when an unrelated
antigen interacted with appropriate naïve CD4 T cells.[109] This and related findings
led investigators to put considerable effort into examining the potential role of the
cytokine environment, in which CD4 T cells are activated, in determining their Th1/
Th2 phenotype.

Two sets of observations were critical. The first involves *in vitro* studies similar
in principle to those with *L monocytogenes,* and the second concerns *in vivo* studies
in mice infected with the protozoan parasite *Leishmania major.* It is found that the
Th1/Th2 phenotype of the CD4 T cells generated when TcR transgenic CD4 T
cells are activated by antigen *in vitro* is greatly affected by the presence of certain
cytokines. Thus, when immunologists deliberately ensure the presence of IL-4 by

adding it to such cultures, the generation of IL-4-producing Th2 cells is greatly favored.[110] Similarly, the presence of IFN-γ[111] and/or IL-12[109] favors the generation of IFN-γ-producing Th1 cells. These and related observations are highly suggestive and interesting. However, the systems employed in making these observations are somewhat artificial. For example, immunologists add to these cultures substantial amounts of IL-4, IFN-γ or IL-12. The question arises as to whether there is a natural source of these cytokines in such amounts in real *in vivo* situations. If IL-4 and IFN-γ are not delivered to the CD4 T cells during their *in vivo* activation, these observations might just be of incidental interest and not provide an insight into how the Th1/Th2 phenotype of CD4 T cells is in reality determined.

Fortunately, the second set of observations, employing intact mice, leaves little doubt that IL-4 and IFN-γ have a pivotal *in vivo* role in determining the Th1/Th2 phenotype of the immune response. In addition, these further observations provide a vivid example of the physiological importance of the Th1/Th2 phenotype of the immune response following infection.

Studies in murine models of cutaneous leishmaniasis

Human cutaneous leishmaniasis is caused by a protozoan parasite that grows intracellularly in macrophages. An individual becomes infected when a parasite-carrying sand fly lands upon the skin of a person and takes a blood meal, leaving parasite-containing saliva in its place. Such infection leads to a limited lesion in some individuals and the parasite is contained close to the initial site of infection. In other individuals, the parasite is not contained but spreads to distal sites, a process sometimes referred to as metastasis. Similar patterns are seen when mice are injected with parasites. In both humans and mice, an infection, primarily limited to the initial site of infection, is referred to as resistance, and the spread of the parasite to distal sites, leading to significant pathology, is referred to as susceptibility.

This mouse model of human cutaneous leishmaniasis has become a major tool for understanding the importance of the Th1/Th2 phenotype of the immune responses in controlling infection, and in understanding how this phenotype is determined. We shall refer a number of times to studies exploiting this interesting model system.

When James Howard infected different strains of mice with a million parasites, he found that some strains are resistant and others are susceptible. Most interestingly, the resistant strains mount a stable, predominant Th1 response over a period of many weeks. Susceptible mice mount a transient Th1 response that evolves, over a couple of weeks, into a predominant Th2 mode, associated with parasite spread to distal surface sites and, with time, to internal organs and widespread metastasis. The question arising from these observations is whether this association of a predominant Th1 response with resistance, and the association of a Th2 response with susceptibility, is accidental or causal, and therefore of prime significance in understanding how immunity can contain this pathogen.

It has been found that resistant mice, when given at the time of parasite challenge an antibody that binds to and neutralizes the activity of IFN-γ, are no longer resistant; they succumb to the parasite. Moreover, this susceptibility is associated with the generation of a predominant Th2 anti-parasite response. These observations lead to the suggestion that the association of resistance with a predominant Th1 response, and of susceptibility with a predominant Th2 response, are causal. They also show that neutralizing the activity of IFN-γ results in a modulation of the response from a predominant Th1 to a predominant Th2 phenotype. This finding is consistent with the idea that, in the absence of IFN-γ, the inhibition of the multiplication of Th2 cells by IFN-γ is lifted, so they can become dominant.[112]

It has also been found that susceptible mice, when given at the time of parasite challenge an antibody that binds to and neutralizes the activity of IL-4, are no longer susceptible but resist the parasite. Moreover, this resistance is associated with the generation of a predominant Th1, anti-parasite response. These observations are again in accord with the idea that the association of resistance with a predominant Th1 response, and of susceptibility with a predominant Th2 response, are causal. They also show that neutralizing IL-4 activity results in a modulation of the response from a predominant Th2 to a predominant Th1 phenotype. This finding is consistent with the idea that, in the absence of IL-4, the expansion of Th2 cells is inhibited, so they cannot become dominant.[113]

Such observations provide compelling evidence for the central importance of IFN-γ and IL-4 in determining in vivo the Th1/Th2 phenotype of immune responses. These sets of observations have led to two different views, on one of which I shall now elaborate. It has been proposed that the prevalence of such cytokines as IL-12 and IL-4, at the time of infection when anti-parasitic CD4 T cells are first activated, determines whether Th1 or Th2 cells are generated. For example, it appears that IL-4 is required to foster predominant Th2 responses, since the administration of anti-IL-4 antibody prevents the generation of Th2 responses. What could be the source of this IL-4? It is known that cells other than lymphocytes, such as mast cells and their close relatives, basophils, and perhaps some dendritic cells or NK T cells, can produce IL-4. Following this line of reasoning, individuals try to explore what cells of the innate system might provide the IL4 required to support the generation of Th2 cells.

Similarly, investigators try to identify the sources of cytokines, in terms of the cells belonging to the innate system, that might foster the generation of Th1 responses. I refer to this view as the ***Cytokine Milieu Hypothesis.***

As more immunologists found the PAMP/Danger Model for CD4 T cell activation compelling, they entertained the idea that PAMPs not only determine whether CD4 T cells are activated, but that the nature of the PAMPs determine the Th1/Th2 phenotype of the activated CD4 T cells.[114,115] We refer to this as the **PAMP Hypothesis.**

Finally, studies have characterized different DC subsets and their different roles in generating different classes of immunity. Observations have been interpreted as showing that the Th1/Th2 phenotype of a response generated against an antigen depends upon what type of DC presents the antigen to naïve CD4 T cells.[116] We refer to this proposal as the **DC Hypothesis.** We shall later consider these and competing ideas.

The existence of multiple steps in lymphocyte activation and its pertinence to the analysis of the underlying mechanism of activation

We illustrate considerations pertinent to the experimental analysis of the requirements to activate naïve lymphocytes by examining the requirements to activate naïve B cells.

A normal mouse has a high level of diverse antibodies in its blood. This antibody has a half-life of a few days, and so antibody is normally being produced and being destroyed through catabolism all the time, even in mice not immunized by an immunologist. We know that some of these on-going antibody responses are directed against gut flora, and others to degraded self antigens. Assays have been developed to detect single cells producing antibody to a given antigen. When one employs such sensitive assays, one usually finds that there are small, on-going immune responses in adult mature mice to an antigen not administered to the mice by an immunologist. Thus, there are some partially activated B cells in mature individuals to most foreign antigens.[117]

We also find in several cases that the activation of a particular lymphocyte, which occurs over a period of days, consists of a series of steps. In the case of B cells, for example, it appears that an initial interaction with antigen, or its surrogate, antibody to the sIg receptor, results in the proliferation of the B cell. The descendants of this B cell must interact with activated T helper cells in a manner outlined in Figure 29 to obtain an antibody response. It is likely that a B lymphocyte, present in the spleen and partially activated by the two step process just outlined, by exposure to environmental antigens such as gut flora, will produce more antibody when appropriate lymphokines are added to the spleen cell population. This type of observation has been sometimes interpreted in the literature as showing that neither antigen, nor a membrane/membrane interaction between the Th and B cell, is required to stimulate antibody production, only the presence of cytokines. This is a correct inference as strictly stated. When certain lymphokines are added to already partially activated B cells, it can be seen that such exposure results in greater production of antibody. However, it is most likely the case that this observation can only be made because there are partially activated B cells among the spleen cells. What we usually want to know is what the requirements are to activate a naïve B cell, not previously exposed to antigen, to yield descendants that produce antibody.

We need, to incisively explore what these requirements are, to determine the circumstances that allow antigen to activate a population of naïve B cells. With such a system, the roles of antigen and of membrane/membrane interactions become apparent.

The recognition that there can be a series of distinct steps in the activation of lymphocytes gives rise to further questions. How are the requirements for completing the whole process related to the requirements of individual steps? Is this question important?

Consider again the activation of B cells. It would seem that the T helper cell must act in a manner such that T helper cells specific for a foreign antigen cannot help B cells specific for a non-crossreacting self antigen, otherwise responses to foreign antigens would continually undermine the inactivation of B cells specific for self antigens. Thus it seems, in terms of the Two Signal version of the antigen-bridge model of the B cell/T helper cell interaction, that the B cell receptor and the T cell receptor must recognize epitopes on the same antigen molecule, as depicted in Figure 16 of Chapter 3. In the context of this Two Signal Model of the B cell/T helper cell interaction, we refer to this requirement as the requirement for the *recognition of linked epitopes* by the receptors of the B cell and of the T helper cell. In terms of the MHC-restricted model of the B cell/T helper cell interaction, see Figure 27 of this chapter, we must reformulate what this requirement means. The B cell receptor recognizes an antigen that is consequently taken into the B cell by a process called *endocytosis*, processed, and the resulting peptides bind newly synthesized class II MHC molecules, directed to the endocytic compartment. The peptide/MHC complexes are then expressed on the surface of the B cell, where they can now be recognized by the receptor of the activated T helper cell. Reflecting on these steps allows us to understand why T helper cells do not act over long distances; effective cooperation requires the B cell's surface Ig receptor to recognize the nominal antigen for which the cooperating T helper cell is specific.

These considerations provide a context for examining how the nature of the overall process of activation of a naïve B cell, involving a series of distinct steps, depends upon the nature of the individual steps. Consider again the activation of a naïve B cell in the context of the MHC-restricted model of the B cell/T helper cell interaction depicted in Figure 27. For convenience, we define the initial interaction of the B cell receptor with antigen, the taking in of the antigen by the process of endocytosis, and the subsequent processing of the antigen and the consequential expression of peptide/class II MHC complexes on the B cell's surface, as step 1. The activated T helper cell then binds this peptide/MHC complex, and its CD40L interacts with the CD40 molecules expressed on the B cell's surface to generate inside the B cell other signals, a collection of processes we call step 2. The T helper cell also delivers lymphokines to the B cell, resulting in various further signals, processes that, for the purposes of this discussion, we designate as step 3. The events involved in step 2 and in step 3 will partially overlap in time, but this does not adversely affect our considerations.

In overwhelming parasite infections, much parasite-specific antibody is produced. The isotype of this anti-parasite antibody depends upon the Th1/Th2 phenotype of the anti-parasite Th cells generated. Moreover, it is found in some overwhelmingly large infections, associated with intense immune responses, that there is an increase in antibody of the same isotype as the anti-parasitic antibody, but substantial amounts of this antibody is unable to bind to parasite antigens. How is such antibody produced? We know that there are many **on-going immune responses** in the spleen of a typical mouse. When overwhelming anti-parasite immune responses occur, it is sometimes possible to detect higher levels in the blood of the lymphokines associated with the Th lymphokine profile of the anti-parasite specific Th cells. It seems inevitable that these lymphokines stimulate partially activated B cells, specific for non-parasite antigens, to produce more antibody of the isotypes favoured by these lymphokines. In this way, parasite-specific Th cells can help responses to unrelated antigens. This is surprising, at first glance, as the hallmark of the immune system is its specificity. How is it possible that nature tolerates a situation in which anti-parasite Th cells affect antibody responses to non-crossreacting antigens? This would be a particularly worrying phenomenon if the anti-parasitic immune response gave rise to the production of antibody against peripheral self antigens. i.e. to autoantibodies.

Let us consider again an elemental situation. Consider a naïve B cell that has to go through all three steps, outlined above, to yield descendants that produce copious amounts of antibody. We have discussed the fact that step 2 requires the operational recognition of linked epitopes by the B cell and Th cell. Such linked recognition does not seem to be strictly required for step 3, though the response by the partially activated B cell upon exposure to the Th cell-produced lymphokines, via step 3, is likely much more efficient if linked recognition occurs. However, if all these steps are required to activate naïve B cells efficiently, then the B cells specific for non-parasite antigens, which respond to the lymphokines produced by parasite-specific Th cells, will have gone through step 2. We can describe the situation by saying that step 2 requires the operational recognition of linked epitopes, and that the requirement for such recognition is somewhat leaky for step 3. However, those B cells that have not gone through step 2 are not receptive to the signals involved in step 3. These considerations illustrate the general rule that to fully activate a naïve cell through several mandatory steps, the most stringent step determines the stringency of the overall process. The stringency of step 2 means it acts as a checkpoint.

I use a holiday analogy for my students. Suppose you want to take a load through a Greek village on a donkey in the heat of summer. If wise, you will examine the path through the village and take note of its narrowest point and not load the donkey in a manner that it cannot traverse it.

In summary, the recognition that there are multiple steps in activating lymphocytes leads to an appreciation of the importance of having precursor cells in a ground state, that have not gone through some of the early steps of activation, if one wants to ascertain the full requirements for activating naïve lymphocytes. In addition, although some steps may not strictly involve certain requirements, such as the recognition of linked epitopes, the stringency of completing the whole process will reflect a requirement if just one of the obligatory steps involves the requirement. For example, the overall process will require the operational recognition of linked epitopes if one step requires such recognition.

A further branch of the humoral arm of the immune system: limitations of the Th1/Th2 paradigm and the existence of anti-inflammatory immunity

We have already noted that the Th1/Th2 paradigm is an oversimplification. Different classes and subclasses of antibody are not induced under the same conditions, and so their induction cannot similarly follow from the generation of Th2 cells. We have seen that individuals from the same geographical region, some allergic to an environmental allergen and others non-allergic, are both humorally immune to the allergen, but the isotypes of the anti-allergen antibodies differ between these two types of individuals. According to one recent study, individuals allergic to an airborne allergen always have IgE antibody specific for the allergen, as well as substantial amounts of IgG_1 antibody, and often some IgA and IgG_4 antibody. Non-alleric individuals in contrast have relatively low levels of IgG_1 and IgE antibody but more substantial levels of IgA and IgG_4 antibodies.[3] These observations raise a number of questions. Are there any untoward consequences for the asymptomatics in having such immunity? What advantage is there in the ability to produce these different classes of antibody, and what circumstances determine which classes of antibody are produced?

Pivotal clues to answering these questions are provided by both older and recent findings. Older observations show that IgA is often produced following chronic antigen exposure via *mucosal surfaces*, i.e. exposure of the body from the outside in a manner not necessarily involving damage, such as occurs via the gut,[118] the urinary and genital tracts, or through inhalation. These IgA antibodies are notoriously ineffective in activating effector functions, such as complement mediated lysis or enhancing phagocytosis. Much of the IgA antibody produced is delivered to the mucosal surfaces upon which the antigen impinged. One of the recognized roles of IgA antibody is to bind to invaders and to thereby minimize or prevent their penetration of the mucosal surfaces. The more recent findings concern the molecular nature of human IgG_4 antibody and how the production of it and IgA antibody is regulated.

States of anti-inflammatory immunity

Human IgA and IgG_4 antibodies do not activate effector functions, but can block the activity of other Ig classes of antibody in their initiation of an attack on the entities these antibodies recognize. It has been recently established that IgG_4 antibodies, produced as divalent molecules when synthesized inside a cell, readily split apart when in the blood into monovalent halves and recombine with other half IgG_4 molecules to become bi-specific, and thus univalent for the antigen that originally led to their production. These recombined IgG_4 molecules are potent in blocking the effects of other classes of IgG and IgE antibody in activating effector mechanisms by virtue of their ability to inhibit the formation of substantial complexes of other antibodies with their antigen.[119] Moreover, we have seen that non-allergic individuals living in geographical areas where an allergy is prevalent produce relatively little IgG_1 and IgE antibody specific for the allergen but produce substantially more IgG_4 and IgA antibody than allergic individuals.[3] Thus, humans can produce both allergen-specific IgG_4 and IgA antibody without the production of substantial amounts of any other known class of antibody. In other cases, the levels of IgE antibody can be similar in all individuals, but the non-allergic individuals can have about a hundred-fold higher level of IgG_4 antibody.[120] It seems appropriate to describe these immune states, present in non-allergic individuals, as a state of *anti-inflammatory immunity*, recognizing the blocking activity of IgA and IgG_4 antibody. Other recent developments support the appropriateness of this designation, as we shall shortly see.

A novel feature of the Theory of Immune Class Regulation, proposed in the early 1970s, was its attempt to provide a meaningful relationship between the different natures of the effector functions of distinct classes of immunity and the different conditions under which these distinct classes are generated.[59] As previously suggested, it seems that cell-mediated and humoral immunity have evolved to provide immunity effective against the foreign invader under most circumstances and, at the same time, to minimize the damaging consequences of any autoreactivity. These considerations, made over forty years ago, were in the context of there being just two main kinds of immunity, namely cell-mediated and humoral. Considerations of a similar kind seem pertinent today in the context of the recognition that different classes of antibody have different abilities to activate effector functions, and even to block such effector functions. Indeed, it makes some sense that chronic exposure to antigen leads under some circumstances to states of anti-inflammatory immunity. Thus chronic, mucosal immune responses caused by gut flora induce IgA antibody, notoriously inefficient in activating any effector functions, and chronic exposure of people to antigen by mucosal and non-mucosal routes can lead to the generation of blocking IgG$_4$ antibody.[120] One type of chronic response that sometimes occurs is that against self antigens. Most self antigens, if attacked, are rapidly replenished, inevitably leading to chronic immune responses. We shall later further consider the significance of such responses to self antigens.

The Significance of T$_{reg}$ cells

A topic of great interest, and currently the subject of intense research, is the nature and function of *regulatory T cells*, or *T$_{reg}$ cells*, as they are often referred to. These are antigen-specific CD4 T cells that produce IL-10 and TGFβ upon stimulation with antigen. They also express various surface antigens, such as CTLA-4, more consistently than do other CD4 T cells. It is unclear how many distinct types of T$_{reg}$ cells there are, but there appear to be at least two. *Natural T$_{reg}$ cells* were the first to be discovered. Sakaguchi and colleagues found that mice, thymectomised when between three and ten days of age,[121] develop various forms of organ-specific autoimmunity some months later. The appearance of this autoimmunity could be prevented by giving the thymectomised mice, shortly after their thymectomy, either CD4+ thymocytes, or peripheral CD4 T cells, harvested from mice older than ten days. These thymocytes or peripheral CD4 T cells contain *natural T$_{reg}$ cells* that suppress the generation of autoimmunity that would otherwise appear in the thymectomised mice. Thymectomy between three and ten days of age results in autoimmunity because the thymus exports a population of T cells deficient in T$_{reg}$ cells from days zero to ten, after which the generation of T$_{reg}$ cells in the thymus, and their emigration to the periphery, is more prevalent. Observations have led to the plausible suggestion that natural T$_{reg}$ cells are biased towards recognizing peripheral self antigens.[122]

It became clear with time that other IL-10- and TGFβ-producing CD4 T cells could be generated in the periphery. It appears that sustained stimulation by antigen at mucosal surfaces, such as through the gastrointestinal tract or through inhalation of antigen, can result in such *induced T$_{reg}$ cells*.[123] It is intriguing to consider the possibility that both natural and inducible T$_{reg}$ cells have a similar function, as might be inferred from their ability to produce the same cytokines, namely IL-10 and TGFβ.

These cytokines in humans respectively favor the production of IgG_4 and IgA blocking antibodies, and can inhibit the production of cytokines by Th1 and Th2 cells.[3,124] Thus, these T_{reg} cells not only help in the production of IgG_4 and IgA antibody, but act to inhibit the function of Th1 and Th2 cells. These facts would seem to reinforce the appropriateness of referring to the predominant production of IgG_4 and IgA antibody, associated with the predominance of T_{reg} cells, as a state of ***anti-inflammatory immunity***. It appears highly unlikely that such states of anti-inflammatory immunity can be generated following acute infections or acute stimulation by antigen, but rather that their generation requires chronic antigen stimulation. The requirement for chronic stimulation would protect the acutely infected host from the deleterious effects of a state of "anti-inflammatory immunity" in dealing with acute infections. We shall later discuss what could be the advantages of such anti-inflammatory immune states.

False trails

In perusing my description of the main events in the field that have occurred since 1970, I realize that I have given in some sense an incorrect impression. This is because I have concentrated on those studies that seem most insightful and that led to clear and acknowledged advances. However, there are always false leads, and some major preoccupations, which do not lead to clear progress and which, to be honest, prevent progress by their dominance of the field. These false trails, if they become fashionable, deviate scientists from investigating more profitable subjects. It is a fact that science is carried out by groups of people who have a shared interest in a particular topic, or who have a common commitment to some broad way of looking at things. Thus the focus of my description on successful advances gives the impression of inexorable progress. This really is not how science unfolds.

Sometimes insights, once gained, are subsequently lost. Some prominent conceptual platforms for enquiry are later found to be invalid. It is important for the non-scientist, who wishes to get a feeling of what it is really like to do science, and youthful idealists, beginning to embark on their scientific career, to recognize how many studies are not so profitable, or are actually misleading. The uninitiated might well think that the large majority of papers represent progress and are valuable to a greater or lesser degree. Indeed, this is the impression often given when the history of the subject is told. Blind alleys are usually ignored, and observations that are not confirmed over time are not described in most historical accounts. Such a distortion of what actually occurs leads to a belief that progress is inexorable. The resulting vantage point fails to recognize the importance of subtle and critical judgment; this failure, in turn, leads to an uncritical mentality that accepts implausible observations and ideas as valid, sometimes providing an accelerating path into an intellectual abyss.

Indeed, doing good science is much more challenging than it would be if all reports represented progress. The worthiness of what you achieve, in terms of real scientific advances, depends so much upon what you discern as likely to be valid, as you survey the whole field. Such a judgment involves weighing the evidence, and assessing the plausibility of the associated ideas against one's own intuition. Such private assessments are critical for a serious investigator, in their choice of a problem worthy of attack and the path chosen for its investigation. This is why I tell my graduate students that having wrong knowledge in their head is much more debilitating to their creative thinking and the research they do than is an absence of correct knowledge. It is mainly for these reasons that I shall try partially to readjust the impression that my description has so far conveyed. I shall briefly describe one or two of the less profitable avenues explored during this time. This description is important for the reasons already indicated, as well as for keeping one's intellectual balance. The reaction to the invalidation of some previously accepted knowledge is often inappropriately severe, leading to the rejection of valuable and potentially valid knowledge. I will illustrate this generality by providing a particular example.

Niels Jerne made several important contributions to immunology, including, as we have seen, the first modern proposal of a selective theory of antibody formation. Jerne proposed some ideas in the early 1970s, referred to as the **Idiotypic Network Theory**,[125] that caught many people's imagination. He suggested that the variable regions of each antibody molecule, as it is produced in greater amounts following the induction of an antibody response, is recognized by the immune system as foreign, leading to the generation of other antibodies recognizing this antibody. Suppose we generically call one of the antibodies, produced in the response to stimulation with the antigen A, the antibody Ab_1. This Ab_1 antibody would, by definition, bind to A. Jerne envisaged that when its production is increased, due to the mounting of an anti-A response, Ab_1 would stimulate the production of antibodies, Ab_2, that recognize the variable regions, or the **idiotypic epitopes**, of Ab_1. This in turn would stimulate the production of Ab_3 molecules, able to recognize the idiotypic epitopes of Ab_2 molecules. These proposed successive interactions between cells, recognizing each other's idiotypes, were the crux of Jerne's **Idiotypic Network Theory**. Jerne's ideas were based in part on the fact that it is possible under certain circumstances to raise anti-idiotypic antibodies. However, these seemed to me at the time, and still seem to me, to be rather special and understandable circumstances.[73]

The ideas Jerne proposed were rather diffuse to my mind. He did not discuss how they might relate to important physiological questions, such as how self-nonself discrimination or immune class regulation might be achieved. In addition, the idiotypic epitopes of an antibody molecule are not synonymous with their binding site.

The defining character as to how an antibody's production should be regulated is, to my mind, most likely defined by how the antibody interacts with self and foreign antigens, rather than by its idiotype. Personally, I could not see a way in which the idiotypic network could serve a physiological purpose when Jerne first proposed his theory, and was doubtful that the network even existed in the broad sense envisaged.[73] I thought the network, in so far as it occurs, would dampen immune responses against foreign antigens and direct the immune system towards unfortunate, internal preoccupations rather than focus on foreign invaders, or aberrant self components, such as cancer cells. I still think that anti-idiotypic responses can occur under some particular conditions and can account for some pathological situations.

Independently of my personal reactions to Jerne's Idiotypic Network Theory, it was obvious that many felt it held promise of providing insight into the inner workings of the immune system. Several of the immunologists I most respected from my reading of the literature embarked upon studies inspired by this theory. A leading laboratory, that of Baruj Benacceraf, devoted themselves to such studies. Benacerraf and colleagues had made significant contributions to the genetics of the ability to produce antibody responses to particular antigens, part of the work cited when Benacerraf received the Nobel Prize in 1980.[126] With time, at least twenty papers were published by Benacerraf's group on a series of T cells that suppress the induction of delayed-type hypersensitivity. The T_s1, T_s2 and T_s3 cells formed a cascade of interacting cells.[127] The receptor of a cell higher in the cascade recognized the idiotype of the receptor of its closest but lower neighbour. Thus, T_s2 cells recognized the idiotype of the receptor of T_s1 cells. The T_s1 cells themselves recognized epitopes present on the receptors of the T cells mediating DTH. This idiotypic epitope was defined by the *variable region of those antibody molecules* that had similar specificity as the T cells mediating the DTH. The observations reported were only readily understandable within the framework that T cells use the same V, D, and J genetic elements, as are used to generate antibody genes, to generate their antigen-specific receptors, and if the receptors of these T cells recognize intact antigen.

Most of the studies on the T_s1, T_s2, and T_s3 cell cascade were published in the *Journal of Experimental Medicine*, a prestigious scientific monthly. There was a time when I collected a stack of these papers beside my bed and tried, at times, to make sense of it all, but to little avail. However, these ideas were so dominant in the field that they were described in immunological textbooks, and I felt compelled for many years to outline them in my undergraduate course in immunology, so that my students had some basis for understanding the contemporary literature.

In time, people realized that most T cells do not directly recognize intact antigen, but rather peptides derived from the nominal antigen bound to MHC molecules, and that the generation of the receptors employed by T cells involve different V, D, and J elements from those employed in generating the genes coding for antibody chains. I realized in time, and with considerable relief, that I could withdraw my description of the unlikely T_s1, T_s2, and T_s3 cell cascade from my undergraduate course.

Jerne's Network Theory was highly cited when Jerne received his Nobel Prize in 1984,[128] probably based in considerable measure on the papers emanating from Benacerraf's laboratory.[127] When the T_s1, T_s2, and T_s3 cell cascade became implausible, its description disappeared from standard texts. There was little or no public discussion among immunologists as to how such a body of observations came to be reported and accepted in the field. I find it unfortunate that Jerne's contribution of the Idiotypic Network Theory was so prominently cited as grounds for the award of the Nobel Prize, as Jerne made other, more significant contributions to the field.

There were also many reports in the 1970s and 1980s on "suppressor T cells" from laboratories from every country in which there were immunologists. I do not wish to cover this work even in outline, except to indicate how prevalent were reports that most would now regard as suspect, and to also indicate my own view: with the recognition that some of the observations are likely invalid, the immunological community may have thrown out a baby with the bath water.

The reports of those times, which now seem likely to be invalid, included descriptions of antigen-specific factors produced by suppressor T cells as well as by helper T cells. Such findings were reported in such eminent journals as *Nature*. There were even reports of antigen-specific factors, composed of two different chains, both with variable parts, each chain being made by a different cell. Such findings are hard to reconcile with the Clonal Selection Theory. Some of these suppressor T cells and the factors they made were held to contain epitopes, the so-called I-J epitopes, recognizable by certain antisera. These antisera were made by reciprocal immunization of mice that were genetically identical except for the MHC. It was natural to suppose that the MHC region contained a locus, the I-J locus, at which the two strains had different alleles, say I-J^a and I-J^b, so that reciprocal immunization led to the production of anti-I-J^a and anti-I-J^b antibodies. However, when the DNA of the two strains was sequenced in the pertinent region, the only differences found could be accounted for by known genes whose products were serologically different from what the anti-I-J antisera recognized. It was inferred that the I-J locus does not exist. There had been multiple claims that these I-J epitopes were found on at least some CD8 TsAb cells. All the studies describing the existence and properties of CD8 TsAb cells became suspect. It became recognized in time that the submission of papers in which the word suppressor T cell was employed would almost inevitably be rejected. I cannot go into all the complexities surrounding these issues. However, I wish to state that we found CD8 TsAb cells in mice locked into a cell-mediated mode, as already described. We have also described how the four different conditions under which CD8 TsAb cells are generated are the same four conditions under which DTH is generated, and under which the immune response can be locked into a cell-mediated mode, and that these same four conditions can all be accounted for by the Threshold Hypothesis. I strongly believe in the existence and importance of these CD8+ TsAb cells. However, though once the subject of literally hundreds of papers, CD8+ TsAb cells are now, for the most part, persona non-grata. Despite this general rejection, a substantial role for CD8 T cells in favoring Th1 responses exists in the literature on immune responses to MHC and other antigens, as I have recently reviewed elsewhere.[60]

Synopsis of Chapter 5

The Major Histocompatibility Complex (MHC): Its Discovery and Characterization

Transplantation studies provided the first evidence of the MHC. It was found that primary grafts, typically of skin between different strains of mice, were rejected either within fourteen days or took longer than twenty days. Genetic studies showed that the more rapid rate of rejection occurred when there were genetic differences in the MHC between the donor and recipient mice.

Reciprocal immunization with spleen cells of MHC-incompatible mice led to the production of antibodies that recognize MHC antigens. These antibodies were used over decades to chemically isolate MHC antigens, and to analyze their chemical structure. This chemical purification also lead to a characterization of their three dimensional structure by X-ray analysis. Such antisera were also used to study the inheritance of these molecules, and to construct a genetic map of the MHC.

There are two kinds of MHC molecules. Class I MHC molecules exist on all nucleated cells. They consist of two chains. One chain, β_2-microglobulin, is virtually invariant. The other is highly variable within the species, and in the mouse is coded for at one of two genetic loci of the MHC, denoted as K and D. There are remarkably many different alleles at these two loci. Moreover, these alleles code for polypeptide chains that have different amino acids at many different positions of the polypeptide chain. Thus, most wild type mice will have two different class I MHC molecules containing K polypeptide chains and two containing D polypeptide chains, as they usually inherit different alleles from each parent at both genetic loci.

Class II MHC molecules are found on specialized cells that have mechanisms to take in external antigens. Thus macrophages that phagocytose antigens, and B cells that can take up antigen when the antigen binds to and aggregates their antigen-specific Ig receptors, both express class II MHC molecules. The class II MHC molecules on macrophages, dendritic cells, and B cells consist of two variable chains, the α and the β chains. There are two kinds of class II MHC molecules in the mouse, I-A and I-E molecules, respectively *consisting of A_α and A_β, and E_α and E_β chains, coded for by genes at the A_α, A_β, E_α, and E_β loci of the MHC. There are again many different alleles at these loci in different mice, and these alleles code for polypeptide chains that have different amino acids at many different positions of their polypeptide chain.*

Class I MHC function and the recognition of antigen by CTL

Class I MHC molecules play a central role in a mechanism by which the immune system monitors the internal components of cells. Samples of proteins synthesized inside the cell are cut up in a structure called the proteasome, and some of the peptides generated bind to the grooves of newly synthesized class I MHC molecules. These loaded class I MHC molecules are subsequently expressed on the cell's surface. In this manner, the immune system gains information on the internal proteins of a cell. Some of the class I MHC molecules of a virally-infected cell will present peptides derived from viral proteins, and these viral peptides will not be presented by the class I MHC molecules of uninfected cells. Upon viral infection, a train of events occurs that results in the generation of CD8 CTL, which recognize viral peptide/class I MHC complexes on the virally-infected cells, and consequently lyse these infected cells, thereby minimizing viral replication. The critical, initial observations, that led to this understanding were made by Peter Doherty and Rolf Zinkernagel, when they demonstrated that CTL recognize something derived from the virus as well as class I MHC molecules on the infected cell.

The insight that a CD8 T cell recognizes a peptide bound to the groove of a class I MHC molecule sheds light on old puzzles, such as the biological significance of the diversity of MHC molecules. The diversity of class I MHC molecules, present within a species, is focused around the peptide-binding groove. Thus different class I MHC molecules bind different peptides. The diversity of MHC molecules makes it difficult for a virus or other intracellular parasite to evolve in a manner so that none of the peptides, generated from its proteins, bind to any of the class I MHC molecules present in the population. The diversity of MHC molecules appears to guarantee a virus' failure to realize such a means of evasion of immune surveillance upon infection of diverse individuals.

Antigen processing, presentation and the function of class II MHC molecules

On exposure to the priming antigen, primed CD4 T cells are stimulated to divide and produce lymphokines in the presence of macrophages or other antigen presenting cells. This system provides a rather simple means of assessing under what circumstances T cells recognize antigen. It was found that both intact antigen, or protease-fragmented antigen, could stimulate primed CD4 T cells. This contrasted with the finding that primed B cells respond much better to the antigen they were primed against than to protease-fragmented antigen. Moreover, antigen-pulsed macrophages can stimulate primed and purified CD4 T cells to proliferate and produce lymphokines without any other source of antigen.

The events that allow macrophages to provide antigen in a form capable of stimulating T cells can be split into two main steps. The first process is called **antigen processing.** *This involves the uptake of external antigen by a mechanism such as opsonization, the cutting up of the protein antigen by proteases into oligopeptides of about 10-20 amino acids in length, and the binding of a sample of these oligopeptides to the grooves of newly synthesized class II MHC molecules. These peptide-loaded class II MHC molecules are then inserted into the external membrane of the cell, where they can partake in the second process called* **antigen presentation.** *A clue to the nature of antigen presentation came from the observation that macrophages, not previously exposed to OVA, can stimulate CD4 T cells primed to OVA if incubated with a protease digest of OVA. Such observations led to the proposal that processing leads to the generation of OVA-peptides bound to the surface of the macrophage. All antigen presenting cells (APC) express class II MHC molecules on their surface. It was natural to suppose that the OVA peptides might be presented to the OVA-primed CD4 T cells, bound to the grooves of class II MHC molecules. This possibility would explain why a successful interaction between antigen-primed CD4 T cells and antigen-pulsed APC was class II MHC restricted. This possibility was also strongly supported by the finding that the interaction between antigen-pulsed APC and CD4 T cells could be blocked by antibody to class II MHC molecules.*

Definitive evidence for this model came from another direction. Some strains of animals will generate immune responses when immunized with simple, foreign molecules, such as poly-L-lysine (PLL), whereas other strains will not. The ability to respond appears to be governed by whether peptides derived from the antigen can bind to the animal's class II MHC molecules, whereas the same peptides will not bind to the class II MHC molecule of non-responder animals. This one-to-one correspondence, between the ability of peptides to bind to class II MHC molecules and the ability of the antigen, from which the peptides are derived, to generate immune responses, cannot be fortuitous. These observations provide compelling evidence that whether the class II MHC molecules bind processed antigen is often the critical factor that determines whether an immune response occurs.

Implications of MHC-restricted recognition of antigen by T cells for the interaction between B cells and T helper cells

There are inevitably consequences of such insights as to how CD4 T cells recognize antigen. The Antigen-Bridge Model of the B cell/T helper cell interaction was no longer credible, as T helper cells do not bind intact, native antigen, as envisaged in this model. The MHC-restricted model replaced the Antigen-Bridge Model of the B cell/T helper cell interaction. Hapten-specific B cells were envisaged to endocytose the hapten-carrier conjugate, following the conjugate's binding to and aggregation of the B cell's immunoglobulin receptors, see Figure 27. The receptors and the hapten carrier conjugate they bind land up in a compartment of the B cell, the endocytic compartment, where they are degraded. Newly synthesized class II MHC molecules are directed to this compartment, where they pick up groove-binding peptides, generated by the degradation of the carrier or of other proteins, and the loaded class II MHC are transported to and inserted into the B cell's external membrane. The TcR of an activated helper T cell can then interact with the carrier-peptide/class II MHC complexes, so initiating the interaction between the T helper and the B cell, see Figure 27.

The GOD of Antibody Genes

Tonegawa first showed that the DNA encoding a light chain was in a different configuration in antibody producing cells and in non-lymphoid cells. To obtain a light chain gene, one DNA joining event is needed to bring one of a large bank of V_l genetic elements next to one of a few J_l elements. The generation of one heavy chain gene involves two DNA joining events: the first bringing one of a large bank of V_h genetic elements next to one of several D_h elements, and the second joining event bringing the $V_h D_h$ unit next to one of several J_h elements. The multiple possibilities available give rise to different genes being assembled in different cells. In addition, there are a number of additional mechanisms of introducing variability, not outlined here. This basic combinatorial mechanism has allowed immunologists to understand how a mouse's immune system can code for upwards of 1,000,000,000 different antibody molecules.

A critical aspect of this process is that most attempts at the generation of an Ig gene do not code for a functional light or heavy chain. In this case, further attempts occur. The cell in which these events take place has mechanisms so that, when a gene is generated that codes for a functional chain, for example a functional heavy chain, further attempts to generate heavy chain genes are halted. These regulatory processes result either in a cell not making genes coding for functional antibody chains, in which case the cell dies, or in making a set of genes that code for one functional heavy and one functional light chain, in which case the cell produces just one kind of functional antibody molecule. Thus, studies in the late 1970s and 1980s led to a reasonably detailed understanding of how the vision of the formulators of the Clonal Section Theory is realized at the molecular level, namely how a precursor cell manages to express just one receptor.

The Generation and Use of Transgenic mice

Biologists have found the means for inserting extra DNA containing genes into the chromosomes of mice. They are called transgenic mice and the extra gene or genes are called transgenes. Usually, the DNA containing the gene also contains important DNA elements that control the expression of the gene, so it is expressed in the same tissues where its product is normally found.

A group of transgenic mice of particular interest to immunologists are those transgenic for genes that code for rearranged and functional light and heavy immunoglobulin chains. In this case, developing B cells find they have a functional light and heavy chain, and so the mechanism to **switch off** *the rearrangement of the V, D and J elements is* **switched on***. All the B cells of such a transgenic mouse express the transgenes encoding the functional light and heavy chains. Normally, B cells specific for a particular antigen are scarce, and so it is difficult to study their properties, such as the requirements for their activation. Mice transgenic for genes encoding a functional antibody specific for a known antigen are invaluable as a ready source of B cells specific for this antigen.*

T cells clones and B cell hybridomas

On antigen stimulation, some activated T cells produce IL-2, an apparently universal growth factor for activated T cells. People found that they could grow single activated T cells if they stimulated them with antigen, APC and IL-2. Populations of such T cells, grown from one T cell and so all having the same T cell receptor, are called T cell clones. It is possible to rest the cells of T cell clones and to then examine the requirements to activate them. Such studies have provided clues as to the requirements to activate normal T cells, see the paragraphs under the tiltle "The Activation of CD4 T cells" a couple of pages further on.

Kohler and Milstein found a way of fusing an antibody-producing cell and a myleoma cell, a cancerous antibody-producing cell, stably together. The hybrid cell, referred to as a B cell hybridoma, produces the antibody of the antibody-producing parental cell and inherits the immortality from its cancerous parent. Cells of such hybridomas thus continuously multiply and produce antibody in large amounts. Such antibodies, chemically homogenous, are called monoclonal antibodies. They have become essential reagents in much research and in clinical settings.

Isolating the T cell receptor

Mark Davis developed an ingenious approach to clone genes expressed, i.e. transcribed into mRNA and presumably translated into polypeptide chains, in T cells but not in B cells. He managed in this way to find genes that code for the α and β chains of the T cell receptor. Further studies led people to demonstrate that similar but different V, D, and J elements, from those involved in the generation of immunoglobulin genes, are involved in the generation of the TcR α and β chains. However, the overall rearrangement process involved in the generation of the α and β genes of the T cell receptor is similar to that involved in the generation of immunoglobulin genes. Successful attempts to generate functional TcR α and β chains result in a suppression of further attempts at rearrangement. It is therefore possible to create mice transgenic for rearranged α and β genes, encoding a particular T cell receptor. All the T cells of such transgenic mouse will express this transgenic TcR.

The generation of T cells

T cells are generated in the thymus. There are at least three ways to death for a developing T cell in the thymus. First, pre-T cells, which are the immediate parents of T cells, start rearranging their different TcR genetic elements in attempts to generate genes coding for functional α and β chains of the T cell receptor. If these attempts have been exhausted and are unsuccessful, the cell dies. Second, the rearrangements may be successful in producing a functional TcR, but this TcR may recognize a self antigen that is well expressed in the thymus, as most self antigens are. In this case, the interaction of the TcR with its peptide/MHC complex inevitably leads to its death; this interaction is exactly as envisaged by Lederberg and articulated in his 1959 Science paper.[34] This deletion of anti-self T cells is referred to as **central tolerance**. *Third, all T cells, even those with a functional TcR, die unless* **positively selected**; *this positive selection requires the TcR of the developing T cell to engage with the class I or class II molecules of special thymic cells involved in this process. Those developing "double positive" CD4+CD8+ T cells, whose TcR interact with a self class I MHC molecule, become "single positive" CD8 T cells, and those double positives whose TcR interact with a self class II MHC molecule become single positive CD4 T cells. These processes select for T cells whose receptors can bind to self-MHC molecules. These T cells are useful in the periphery in protecting the host. Positive selection of a T cell is sometimes referred to as giving the T cell the kiss of life. The T cells, whose receptors do not bind to self-MHC molecules, die, a process often referred to as death by neglect. Positive selection allows the frequency of useful, antigen-specific T cells to be considerably greater than if all T cells, with functional TcRs, emigrated from the thymus into the periphery, landing up in secondary lymphoid organs.*

Cellular Interactions

B cell/T helper cell interaction

The interaction of a helper T cell with a B cell involves the formation of an immunological synapse between the T cell and the B cell. In addition to the interaction between the TcR and the peptide/MHC complex, this interaction involves an interaction with CD40 on the surface of the B cell and CD40L, the ligand for CD40, on the activated T helper cell, besides many other interactions. The CD40L molecule is much more strongly expressed on activated than naïve T cells. The T helper cell also delivers lymphokines to the B cell. Different kinds of T helper cell produce different spectra of lymphokines. The nature of this spectrum is critical in determining the B cell's fate, and so critical in the nature of the regulatory function of the T helper cell.

The activation/inactivation of B cells and of CD8 cells

The use of mice transgenic for antibody genes has allowed immunologists to more incisively address how antigen activates and inactivates B cells. These studies have supported the proposal that, in the absence of T helper cells, antigen inactivates B cells, whereas these B cells can be activated by antigen in the presence of helper T cells. Other studies similarly show that CD8 T cells are inactivated when they interact with antigen in a sustained manner in the absence of T helper cells, whereas their sustained activation requires CD4 T helper cells specific for the antigen. These studies are in accord with the model of one cell/ multiple cells for the inactivation/activation of lymphocytes. It is clear from these findings that CD4 T helper cells often control the fate of other lymphocytes on their encounter with antigen.

The activation of CD4 T helper cells

All currently discussed models for the activation of CD4 T cells appear somewhat similar, as they are all cast within a two signal framework. Moreover, signal 1 is identical in all models. Signal 1 is envisaged to be generated following the interaction of the T cell's receptors with peptide/class II MHC complexes on an antigen-presenting cell. The sustained generation of signal 1 alone, i.e. in the absence of a critical signal 2, is envisaged to result in the inactivation of the CD4 T cell, causing the death of the CD4 T cell or rendering it refractory to activation. The models differ in how they propose the generation of the critical signal 2, or the costimulatory signal required for activation, is regulated. Sometimes the differences between these hypothetical models seem minor mechanistically, but these differences have profoundly different physiological consequences.

Lafferty and Cunningham tried to understand why MHC antigens generate such strong immune responses. They found a number of instances where, when lymphocytes are stimulated with MHC disparate cells, the vigor of the stimulation did not correlate with the level of expression of the foreign MHC molecules on the stimulating cell. They suggested that good stimulators had two properties: they bore the foreign MHC antigens and in addition had an extra property that resulted in strong stimulation. Cells with this additional property were said to have an S^+ phenotype, see Figure 30.

The Lafferty/Cunningham analysis was in time supported by further observations. These subsequent studies employed cells from a Th1 clone, specific for a peptide p bound to a class II MHC restriction element R, i.e. specific for p/R. Cells to be stimulated were rested and conditions found where they were barely kept alive. It was anticipated that the requirements to activate such cells might resemble the requirement to activate naive CD4 T cells. These rested T cells were allowed to interact with p/R alone, by exposing them to purified R, inserted into a lipid bilayer containing no other proteins, in the presence of p. Quill and Schwartz showed that these rested CD4 T cells did not proliferate or produce cytokines under these circumstances. However, the rested T cells stimulated by p in the presence of APC proliferated and produced cytokines. It appeared more was required to stimulate the rested T cells than signal 1 and so, in terms of the Lafferty/Cunningham model, the observations implied that APC had an S^+ phenotype.

Two further findings made in this system were seminal. Exposure to p/R alone for as little as forty-eight hours made the rested T cells refractory to stimulation with an APC decorated with p. It appeared that signal 1 alone, sustained over a period of about two days, rendered the rested CD4 T cells unresponsive to activation signals. The generation of signal 1 alone thus appears to inactivate CD4 T cells in this system. Secondly, it became apparent with time that the extra-something the APC had, besides the ability to present peptide/MHC complexes, was a molecule on its surface called B7, which interacted with a receptor on the CD4 T cell, called CD28. The interaction between the APC B7 molecule and the CD28 T cell counter receptor constitutes signal 2, or alternatively called the costimulatory (CoS) signal, see Figure 31.

Several further observations supported this model. For example, the addition of molecules that bind to B7 molecules, and consequently block the B7/CD28 interaction, abrogate the ability of APC to present antigen in a manner to stimulate primed CD4 T cells to proliferate and their descendants to produce cytokines. It was thus recognized that APC normally carried out three processes necessary for CD4 T cell activation: processing of the antigen, presentation of processed antigen, and costimulation.

This model for CD4 T cell activation was important for two reasons. It accommodated many observations and also convinced people of the usefulness of a Two Signal Model for describing the activation of CD4 T cells. There were, however, some difficulties. Macrophages express the B7 molecules constitutively, and they ingest, by various means, both peripheral self and foreign antigens. How could the macrophage inactivate the CD4 T cells specific for peripheral self antigens and activate CD4 T cells specific for foreign antigens? It would appear, if the CoS molecules are constitutively expressed, that the macrophage would indiscriminately activate CD4 T cells specific for peripheral self and foreign antigens. This conclusion seems problematic. I refer to this model as the Constitutive Model of CD4 T cell activation as the expression of the CoS molecule on the APC, and believed to be required for CD4 T cell activation, is constitutively expressed. This critical feature distinguishes this model from other models of CD4 T cell activation.

Janeway proposed a new model in 1989. He pointed out that many non-microbial antigens are only immunogenic if administered with adjuvants containing microbial products. He suggested that microbial products are necessary to activate the APC to express sufficient levels of CoS molecules to fully activate CD4 T cells. Janeway thus proposed that the immune system does not distinguish self from nonself at the level of CD4 T cells, but rather non-infectious entities from infectious entities. Pattern recognition receptors (PRR) of the innate defense system can recognize pathogen-associated molecular patterns (PAMPs), and Janeway envisaged that such recognition is required if the APC is to express the CoS molecules in a manner that allows the activation of naïve CD4 T cells, see Figure 32.

An enigma in the context of Janeway's proposal is that foreign vertebrate antigens, bereft of PAMPs, are immunogenic. Such antigens include skin grafts from genetically disparate mice. Polly Matzinger subsequently developed a model related to that of Janeway's. Matzinger attempted to address this enigma. She proposed that the immune system senses danger by various means. The danger signal resulted in maturation of APC so they could activate CD4 T cells. Matzinger's proposal is referred to as the Danger Model.

Janeway's and Matzinger's proposals stimulated an enormous number of interesting studies. Many reports document how the interaction of PRRs with PAMPs facilitate the maturation of dendritic cells, as well as changing the characteristics of other cells. The PAMP/Danger Model of CD4 T cell activation is the predominant model of the day.

The Th1/Th2 paradigm

Coffman and Mosmann propagated CD4 T cell lines by culturing primed lymphocytes with antigen. They then derived CD4 T cell clones from these lines. An examination of the functional properties of these CD4 T clones, as well as the cytokines they produce, led to their classification as Th1 and Th2 clones. Cells of Th1 clones produce IL-2 and IFN-γ, and can mediate DTH reactions, whereas cells from Th2 clones produce IL-4 and IL-5 and were potent T helper cells for the production of some classes of antibody. These generalizations brought a new level of molecular characterization to the biology of CD4 T cells.

A third arm of the immune system

There are two recognized limitations of mapping Th1/Th2 cells onto cell-mediated/antibody responses. First, Th1 cells can make cytokines, IFN-γ in particular, that can aid the production of some antibody, in particular IgG_{2a} antibody in the mouse and IgG_2 antibody in humans. Th2 cells facilitate the production of IgG_1 and IgE antibody. The isotypes of IgG antibody produced against an antigen can thus be used to infer the Th1/Th2 phenotype of the corresponding CD4 T cells. Second, given that there are several different classes and subclasses of antibody, whose production is not coordinately controlled during immune responses, it is clear that different conditions give rise to the production of antibody belonging to these different classes. We should recognize that more modes of response than the Th1 and Th2 modes are likely to be delineated in the future. We know enough now to acknowledge that there is at least a third mode of immunity, reflecting a state of "anti-inflammatory immunity".

Prolonged antigen stimulation at mucosal surfaces, in mice and humans, can result in IgA antibody production in the mouse, and IgA and IgG_4 production in humans. Such stimulation can also lead to systemic unresponsiveness for the induction of other classes of immunity. Non-allergic individuals, living in a geographical area where there is a prevalence of allergy to an antigen, produce substantial IgA and IgG_4 antibody but relatively little IgG_1 and IgE antibody to the antigen, as compared to allergic individuals. Neither

IgA nor IgG$_4$ are efficient in activating any effector functions, such as complement and mast cell degranulation.

Recent studies show that human IgG$_4$ antibody, made as a divalent molecule, splits apart into two halves and recombines to create antibody molecules that are univalent for two different antigens. This form of IgG$_4$ antibody, monovalent for a given antigen, can block the activity of other antibodies, such as IgG$_1$ or IgE antibody specific for the antigen, from triggering effector functions. Moreover, human CD4 T$_{reg}$-like cells, which produce TGFβ and IL-10, appear to help the production of IgA and IgG$_4$ antibodies and to down-regulate the production of cytokines by Th1 and Th2 cells. It seems that this state, with predominant production of IgA and IgG$_4$ antibody and prevalence of T$_{reg}$ cells, is appropriately referred to as an anti-inflammatory immune state. This state tends to occur following chronic stimulation by substantial levels of antigen, particularly via mucosal routes.

Chapter 6

FURTHER THEORETICAL AND EXPERIMENTAL INVESTIGATIONS ON LYMPHOCYTE ACTIVATION AND IMMUNE CLASS REGULATION

The potential relationship between cardinal physiological features of the immune system and the mechanisms of immune regulation: "non-interference" and "independence"

We shall consider two main questions in this chapter: what determines whether antigen activates or inactivates naïve CD4 T cells and, if activation occurs, how is the Th1/Th2 phenotype of the activated Th cells determined? I would like to bring to the fore two considerations at the level of the system, central to my approach in thinking about these two questions, before we discuss mechanisms at the cellular and molecular levels.

First, it seems critical physiologically that the activation of CD4 T cells by a foreign antigen should normally not interfere with the simultaneous inactivation of CD4 T cells by peripheral self antigens. I refer to this cardinal and desirable feature of the activation/inactivation of CD4 T cells as reflecting the ***principle of non-interference***. Why do I stress this point? Most contemporary models for the activation/inactivation of CD4 T cells violate this principle, which is a major reason why I am uncomfortable with them. The model I propose is attractive to me partly because it optimizes non-interference.

The second consideration bears on the mechanisms by which the class of immunity is determined. Clinical observations attest to the fact that the Th1/Th2 phenotype of the immune response, generated against a foreign invader, is frequently of critical importance in determining the outcome of this invasion. For example, containment of HIV-1, the virus responsible for AIDS, and of *Mycobacterium tuber-culosis*, the bacterium responsible for tuberculosis, appear to require a predominant Th1/CTL response, as I shall argue later. The evolution of the immune response into a mixed Th1/Th2 mode is associated with chronic or progressive disease. In contrast, some worm infections, including nematodes, usually rapidly induce and are contained by Th2 immunity. It seems inevitable in "hygienic" industrialized societies, and even more frequently in "non-hygienic", non-industrialized societ-ies, that people are often multiply infected at one time with different pathogens or microorganisms. These suppositions lead me to suggest that the Th1/Th2 pheno-type of simultaneous immune responses, against two non-crossreacting pathogens, must normally be *independently* determined. Consider what would happen if the mechanisms of immune class regulation did not normally conform to this ***prin-ciple of independence***. The process normally leading to a Th1 response against *Mycobacterium tuberculosis* would be affected by a concurrent infection by a worm that is associated with a Th2 response, deviating, for example, the response against the mycobacteria towards a Th2 mode. Such a deviation would undermine the con-tainment of the mycobacteria. In fact, a predominant Th2 response to *Mycobacterium tuberculosis* leads to a rapidly fatal form of the disease, if untreated, known as miliary tuberculosis. Such considerations support the idea that the Th1/Th2 phenotype of concurrent immune responses against non-crossreacting pathogens will usually be independently determined.[60]

We shall see that some envisaged mechanisms of immune class regulation are consistent with independence whereas others are not. I would suggest that those not consistent with the Principle of Independence are unsatisfactory, particularly as there is evidence for independence being a valid concept, as we shall shortly see. In addition, as we have seen in Chapter 5, some exceptional circumstances can occur where independence is not seen—in overwhelming parasite infections, for example. As already indicated, I consider these to be rare and to reflect pathological situations.

Cellular interactions mediated by the recognition of linked epitopes in the context of "non-interference" and "independence"

We have recognized that the MHC-restricted model of the B cell/T helper cell inter-action involves the operational recognition of linked epitopes, i.e. a T helper cell specific for a nominal antigen Q can only facilitate the activation of a B cell specific for a hapten h if h is physically attached to Q. I think this is critical. Why? If the pres-ence of activated Th cells really determines whether antigen activates rather than inactivates B cells, as envisaged in the original Two Signal Model of lymphocyte activation and in its more modern versions, how Th cells function has critical patho-physiological consequences. Suppose Th cells do not act in a manner requiring the recognition of linked epitopes. In this case, Th cells specific for a foreign antigen F could and sometimes would, in the presence of F, prevent a peripheral self antigen from inactivating its specific B cells, leading to their activation and so to the production of autoreactive antibodies. The requirement for the operational recognition of linked epitopes is a means of ensuring **non-interference** of the inactivation of B cells by events associated with immune responses to non-crossreacting antigens. I shall argue later that the concept of non-interference is even more important in considering the activation and inactivation of CD4 T cells.

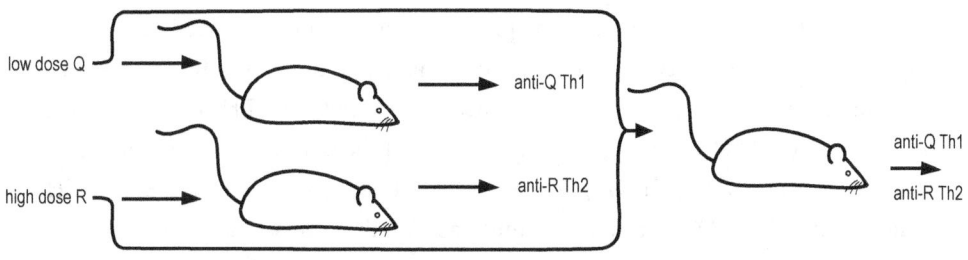

Fig. 33. Demonstration of independence: the Th1/Th2 phenotype of simultaneous immune responses to two non-crossreacting antigens are determined independently

We undertook experiments to critically test the idea that the immune system displays the attribute of **independence**. We found conditions under which immunization of mice, following intravenous injection with the antigen Q, generated a predominant Th1 response in their spleen. We defined other conditions where immunization of mice with the non-cross-reacting antigen R generated, upon intravenous injection, a predominant Th2 response in their spleen. We then injected mice intravenously with both R and Q, mixed together and delivered in the same syringe. The anti-Q and anti-R immune responses seen in the spleen of these doubly immunized mice were, as far as we could tell, identical to those generated in singly immunized mice.[129] Figure 33 provides the outline and summary of the observations of one such experiment. These and related findings lead me to suggest that the immune system, most of the time, displays the feature of independence. Does this feature tell us anything about how CD4 T cells must interact in determining the Th1/Th2 phenotype of the immune response?

We have seen that, once mice are immunized to produce a strong antibody response to an antigen, they are unable to be immunized to express DTH; they are in a state of humoral immune deviation. Such humorally immune mice harbor antigen-specific CD4 T cells that suppress the induction of cell-mediated immunity in the form of DTH, as previously described, see Figure 24, Chapter 4. We showed that these TsDTH cells operationally act by the recognition of linked epitopes. Thus, spleen cells from mice immunized to produce a strong antibody response to the protein *haemocyanin (Hm)* can inhibit the DTH response to *horse RBC (HRBC)* when mice are immunized with the conjugate Hm-HRBC, but not when both Hm and HRBC are present but uncoupled from one another,[79] in the form of HRBC and Hm coupled to mouse red blood cells (MRBC), see Figure 34. In other words, these Hm-specific TsDTH cells act by the operational recognition of linked epitopes. If these TsDTH cells did not act in this way, a humoral Th2 response to one antigen would inhibit a concurrent DTH, Th1 response to another non-cross-reacting antigen, when both antigens are present. For example, a Th2 response to a nematode would deviate the immune response to *Mycobacterium tuberculosis* from a Th1 towards a Th2 mode. In summary, the phenomenon of independence in determining the Th1/Th2 phenotype of concurrent immune responses is critical to the well functioning of the immune system; to realize this attribute TsDTH cells must negatively regulate DTH responses via the operational recognition of linked epitopes. Thus, we have seen that Th cells, in helping the activation of B cells, and of TsDTH cells, in inhibiting DTH responses, must act by the operational recognition of linked epitopes if their activity is to be delivered in an appropriately specific fashion. Analogous considerations allow us to appreciate the importance of such operational recognition of linked epitopes, by all antigen-specific lymphocytes, if they are to act specifically to help or inhibit the induction or activity of other specific lymphocytes.

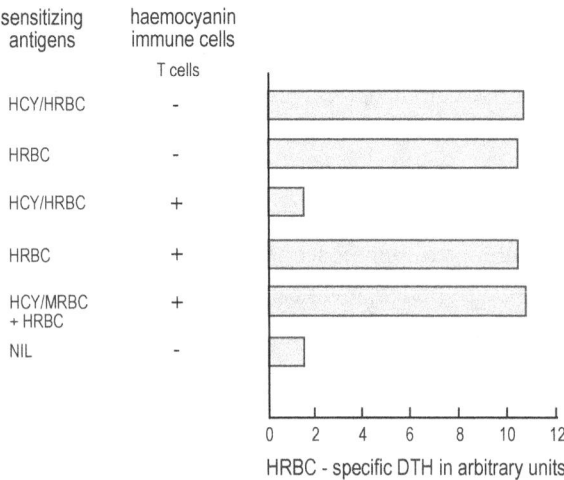

Fig. 34. The TsDTH cells generated during the course of an antibody response suppress the induction of DTH by a mechanism requiring the recognition of linked epitopes. Adapted from reference 79.

The critical importance of the operational recognition of linked epitopes in the activation of CD4 T cells

I now turn to considerations that were critical in my formulation of the model for CD4 T cell activation,[130] published in 1999 and consonant with my current views. I thought about this model for many years. I initially felt it to be somewhat extravagant and therefore implausible; in time, I felt it was the only solution I could think of to a critical problem that preoccupied me. Following Sherlock Holmes' dictum, I decided to take the plunge and suggest that what initially seemed improbable was, indeed, not only a possible solution but a likely one, as all other solutions seemed less plausible. Before describing this model, I shall outline the considerations that led me to this scenario.

When the Two Signal Model of lymphocyte activation was originally proposed, we assumed that all lymphocytes had antigen receptors that recognize intact antigen, just as antibody does. This assumption allows one to readily imagine mechanisms by which CD4 T cell collaboration can occur by a process involving the simultaneous recognition of linked epitopes. For example, we can just adapt the general description of the Two Signal of Model depicted in Figure 16 of Chapter 3 by replacing the precursor cell with a precursor T helper cell.

I personally experienced a theoretical, not to say emotional, crisis when the MHC-restricted nature of the recognition of antigen by T cells became apparent, primarily through the work of Doherty and Zinkernagel. The operational recognition of linked epitopes was essential to the ideas I entertained as to how Th cells could effectively control the activation/inactivation of other lymphocytes, particularly of precursor T helper cells. It seemed essential that the presence of a foreign antigen, F, and activated CD4 T cells specific for F, could not act to subvert the inactivation by a peripheral self antigen S of a single anti-S CD4 T cell. We have already dubbed this desirable lack of subversion as non-interference of CD4 T cell inactivation by S, following the simultaneous occurrence of immune responses to other, non-cross-reacting, foreign antigens.

It appeared that such subversion would at least sometimes occur unless the CD4 T cell cooperation, needed to prevent CD4 T cell inactivation and resulting in CD4 T cell activation, required the operational recognition of linked epitopes. However, when it became apparent that CD4 T cells do not recognize intact antigen, but rather peptides derived from the nominal antigen and recognized in the context of class II MHC molecules, I could not, for a period of several years, envisage how the operational recognition of linked epitopes could occur. How could linked recognition occur if the TcRs of the interacting CD4 T cells recognized different and unlinked peptides derived, by processing, from the same nominal antigen?

I recognized in time that the dilemma I faced bore significant similarities to another of the field. Most immunologists had recognized that the activation of hapten-specific B cells, facilitated by carrier-specific Th cells, required the hapten to be linked to the carrier. Moreover, the physiological importance of this requirement for the recognition of linked epitopes was apparent, as discussed in Chapter 5. How could there be such a requirement for linked recognition when the hapten-carrier conjugate had to be degraded into peptides before the Th cell could recognize the peptide/class II MHC complex?

In considering this question, I became aware that there are two different means by which the operational recognition of linkeds epitopes can be realized. The most obvious is the one in which the two epitopes are simultaneously recognized, with such simultaneous recognition leading to the generation and delivery of a signal that would not occur without such simultaneous recognition, as illustrated in Figure 16. The second means of realizing the operational recognition of linked epitopes is by sequential recognition of epitopes on or derived from the antigen. A receptor recognizes the first epitope, leading to events such that the second epitope is only generated, and so available, as a consequence of this first recognition. Consider, for example, the hapten-specific B cell of Figure 27 (Chapter 5) exposed to the hapten-carrier conjugate h-C. The receptor of the h-specific B cell recognizes h-C, the first recognition event, leading to the internalization of h-C into the cell, its processing into peptides, including the generation of the peptide "c", in an endocytic compartment of the cell. Newly synthesised class II MHC molecules are directed to this compartment and pick up the peptide "c", and the peptide/class II MHC complexes are then inserted into the surface membrane of the B cell, where they can be recognized by the Th cell specific for the nominal antigen C. The peptide "c" would not exist, complexed with the B cell's class II MHC molecules on the surface of the B cell, unless the hapten was coupled to the carrier C. Efficient cooperation of the kind depicted in Figure 27 thus requires, as we have previously seen, the operational recognition of linked epitopes.

Ian Ramshaw and I had found that the CD4 T cells, generated during the course of an antibody response and able to inhibit the induction of DTH, act by the operational recognition of linked epitopes, as already outlined, see Figure 34. I remember Ian and I discussing, probably in the mid 1980s, how this requirement for linked recognition might be achieved if CD4 T cells recognized peptide/MHC complexes. We discussed the possibility that the operational recognition of linked epitopes by the two CD4 T cells could occur if this recognition was mediated by an antigen-specific B cell presenting the antigen. An antigen-specific B cell, recognizing either Hm or HRBC via its Ig receptors, could lead to the uptake of the Hm-HRBC conjugate and to the presentation by this B cell of peptides, some derived from Hm and others from HRBC. In this case, the B cell could mediate the interaction between

Hm-specific TsDTH cells and HRBC-specific pDTH T cells, i.e. T cells able to be activated by HRBC to give rise to DTH-mediating T cells.

These considerations can be summarized thus. To achieve such attributes as independence and non-interference, it is essential that the underlying interactions between T cells are mediated by the operational recognition of linked epitopes. If these T cell interactions involve a CD4 T cell, and as class II MHC molecules are not expressed on mouse CD4 T cells, the interaction between the CD4 T cell and other T cells must be mediated by an APC. It would appear that the operational recognition of linked epitopes can only be ensured if the APC, mediating this interaction, is an antigen-specific B cell. I consider this to be one of the most significant rules, if valid, concerning how immune responses are regulated. We shall later see that this rule allows some otherwise strange observations to be undersood.

The Two Step, Two Signal Model for the activation of CD4 T cells

From sometime in the 1980s, I considered how one might reconcile the nature of the recognition of antigen by T cells, revealed by the studies of Doherty, Zinkernagel and others, with the principles underlying our original Two Signal Model. To be honest, my first reaction was to deny the generality of the MHC-restricted nature of T cell recognition, given the problems I perceived if MHC-restricted recognition was a general phenomenon. However, the evidence became overwhelming with time that my defensive attitude could not be indefinitely sustained. Though the MHC-restricted recognition of antigen was first most clearly shown for CD8 CTL recognizing virally infected cells, studies on the nature of antigen processing, and presentation of peptides by APC, left no doubt that CD4 T cell recognition was also MHC restricted, in fact MHC class II restricted. I recognized that most APC take up antigens by relatively non-specific mechanisms. One could envisage that a macrophage could take in a nominal antigen Q and subsequently express on its surface peptides q1 and q2 bound to the grooves of the macrophage's class II MHC molecules. This would then allow the interaction of two CD4 T cells, specific for the nominal antigen Q to bind to the q1/class II MHC and q2/class II MHC complexes on the macrophage's surface.

For example, an activated q2/class II MHC-specific CD4 T cell might interact with a naive q1/class II MHC-specific CD4 T cell via a macrophage acting as the intermediary APC. The interaction of the receptors of the activated CD4 T cell, with their ligands on the APC, might result in the delivery of signals to the macrophage to up-regulate those costimulatory molecules required to activate the naïve, q1/class II MHC-specific CD4 T cell. This possibility allows one to envision how CD4 T cell cooperation could occur. However, a macrophage can readily take in both a peripheral self antigen, S, and a non-crossreacting, foreign antigen F, and present peptides "s" and "f", derived from their nominal antigens and bound to its class II MHC molecules. In this case, the activation of naïve CD4 T cells, specific for S, could be helped by activated CD4 T cells specific for F. This scenario violates the principle of non-interference: CD4 T cells specific for foreign antigens would be able to interfere with the inactivation of CD4 T cells specific for a peripheral self antigen S in the presence of the foreign antigen. I therefore found this possibility implausible.

One can imagine a way of achieving non-interference. We have seen that B cells express class II MHC molecules on their surface and for the most part are specific in their uptake of antigens. Suppose B cells were unique as APC in some way and had functions not shared by other APC. Then the B cell could, via its antigen-specific receptors, take up a nominal antigen Q, and present peptides q1 and q2 bound to its MHC molecules. Then activated CD4 T cells specific for q2/class II MHC complexes could activate the B cell to express costimulatory molecules that allow the B cell to generate the second, costimulatory signal, required to activate a naïve q1/class II MHC-specific CD4 T cell. This possibility, represented by step 2 in the model outlined in Figure 35, was a formal solution to the problem of how CD4 T cell cooperation could be achieved and how, at the same time, *non-interference* could be realized.[130] However, it seemed to me for quite a long time that this possibility was a far stretch.

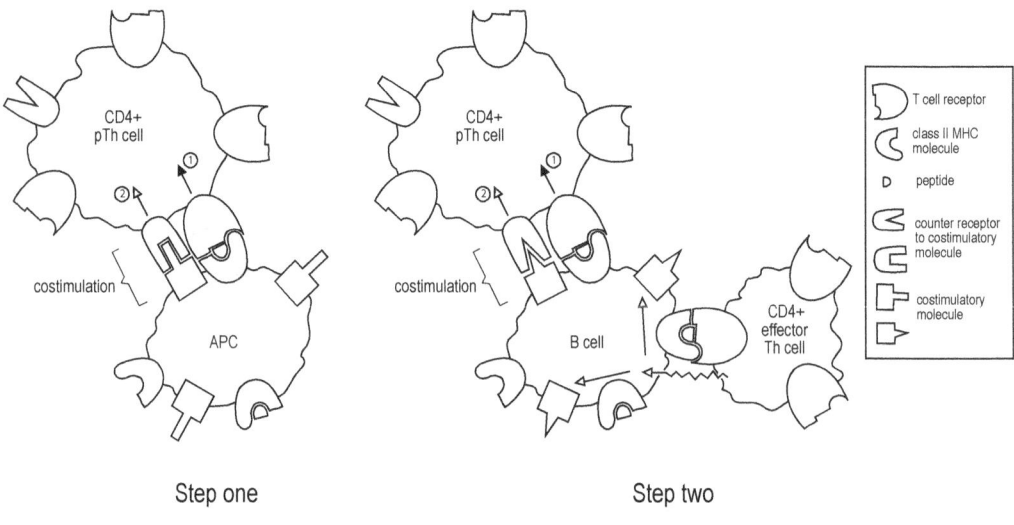

Step one   Step two

Fig. 35. The Two Step, Two Signal Model of CD4 T cell Activation. For explanation, see text.

A major concern of mine when we proposed the original Two Signal Model was the requirement that primary immune responses involved the antigen-mediated interaction between specific lymphocytes that in a naïve mouse are scarce. How could antigen mediate the interaction of these scarce cells?[48] We now know that there are typically a few hundred CD4 T cells specific for a moderately immunogenic antigen among the roughly fifty million CD4 T cells present in a typical mouse.[86] How can these scarce CD4 T cells find each other? I refer to this question as recognizing the **scarcity problem**. The formal solution envisaged above, according to which CD4 T cells might interact in a way consistent with non-interference, involved not just two scarce cells meeting up together, but three: the two interacting CD4 T cells and the B cell specific for the nominal antigen. For years, I regarded this as too extravagant a possibility to be seriously entertained. However, constant puzzling over this question through the years wore down my resistance and led me in the end to consider this formal possibility more seriously. In time, I realized there were five positive features of this potential mechanism. First, there was evidence in the literature, though conflicting, on the need for B cells in the activation of naïve CD4 T cells.[131] Second, naïve B cells express few costimulatory molecules, such as the B7 molecules; however, B cells express these and other costimulatory molecules when activated![132] Why the expression of these CoS molecules should be so regulated if activated B cells had no central role as an APC was a mystery. Third, there were many interesting aspects of the immune system that the basic and simple concepts I had been employing did not take account of. There is a great deal of structure to the lymph nodes and spleen where immune responses are mounted. In addition, we were becoming more aware in the 1990s that there are concentration gradients of molecules called **chemokines** guiding the movement of lymphocytes in lymph nodes and elsewhere.[133] Guidance of movement in such structures might well facilitate the ability of scarce cells to find each other. Fourth, considerable though indirect evidence was consistent with the idea that CD4 T cells could be stimulated by macrophage- or dendritic cell-presented antigen to divide. I imagined this might normally occur as a first step in most CD4 T cell responses, independently of whether this first interaction would lead to activation or inactivation of the CD4 T cells. Such a first step, during which the responding CD4 T cells proliferated, would minimize to a certain extent the scarcity problem. The activation of the step 1 primed CD4 T cells to generate significant levels of cytokine-producing progeny cells was envisaged to require their successful completion of step 2, according to which these step 1 primed CD4 T cells interact with antigen presented by an activated and antigen-specific B cell, see Figure 35. Those CD4 T cells that only go through step 1 are postulated to become inactivated in time, i.e. non-functional. Lastly, there is a quantitative problem if the interaction of the two scarce CD4 T cells is mediated by an APC that is highly prevalent. In this case, it seemed improbable that the two scarce CD4 T cells would engage with the same antigen presenting cell when these antigen presenting cells are exceedingly more prevalent than the CD4 T cells. This "prevalency problem" does not exist if the mediating B cell is an antigen-specific B cell.

This Two Signal Model is again a minimal model. We already know that CD4 T cells will respond to cytokines, such as IL-2, in what could be called a third step. I refer to this model as the ***Two Step, Two Signal Model of CD4 T cell activation***.[130] We shall later in this chapter consider the pertinence of this model in formulating a contemporary version of the Threshold Hypothesis. In addition, we will discuss the pros and cons of the different and contemporary models of CD4 T cell activation and inactivation.

A demonstration that CD4 T cell cooperation facilitates the in-vivo activation of CD4 T cells

A graduate student of mine, Nathan Peters, expressed his wish at the beginning of his graduate studies to test critical predictions of the Two Step, Two Signal Model as his research project. Carl Power, another graduate student, had already modified an assay, described in the literature, in a manner that allowed us to detect single, peptide- or antigen-specific cytokine-producing CD4 T cells among a population of cells primed against the pertinent antigen. This modified ***ELISPOT assay*** has become essential to much of our research as it allows us to enumerate antigen-specific cytokine producing T cells.[134]

Our overall strategy was simple in principle, though not in practice. We sought to find an antigen against which it was very difficult, but just possible, to generate an immune response. We surmised that once T helper cells are activated, other lymphocytes would also be readily activated. We thus also surmised that the difficulty in generating the immune response reflected the difficulty in first activating T helper cells. We further surmised that this difficulty in activating helper T cells was due to the scarcity of helper T cells specific for the nominal antigen, as we envisaged such activation requires CD4 T cell cooperation. We imagined that if we could ablate some of the CD4 T cells specific for the antigen, it would be difficult if not impossible to efficiently activate the remaining CD4 T cells specific for this nominal antigen.

Many studies had been carried out to determine the peptide-specificity of CD4 T cells generated when different strains of mice are immunized with the antigen ***hen egg lysozyme (HEL).*** Most studies then reported used powerful adjuvants, containing microbial products, to generate an immune response to HEL, as HEL is not a very immunogenic antigen. We decided to try to identify the peptide specificity of all the CD4 T cells generated upon immunization with HEL, employing our improved ELISPOT assay. We managed to do so; this was shown by the fact that the sum of the number of CD4 T cells specific for a set of non-overlapping HEL peptides, detected in a population of HEL-immunized spleen cells, was equal to the number detected against the intact HEL protein.[135]

We found it difficult to generate strong antibody responses against HEL in the BALB/c strain of mice without the use of a microbial adjuvant, such as **Complete Freund's Adjuvant (CFA)** that contains dead mycobacteria. We wished to develop an experimental system that did not employ microbial adjuvants, to avoid issues arising from their use. We had to use our ingenuity to obtain substantial responses without employing CFA, using tricks of the trade. We found we could detect substantial HEL-specific IL-2-, IFN-γ- and IL-4-producing CD4 T cells in the spleen of BALB/c mice when we immunized with aggregated HEL given with a non-microbial adjuvant. In all cases, rather more than half the CD4 T cells present were specific for one peptide, the "major HEL peptide", $HEL_{105-120}$, and the rest were specific for other peptides.

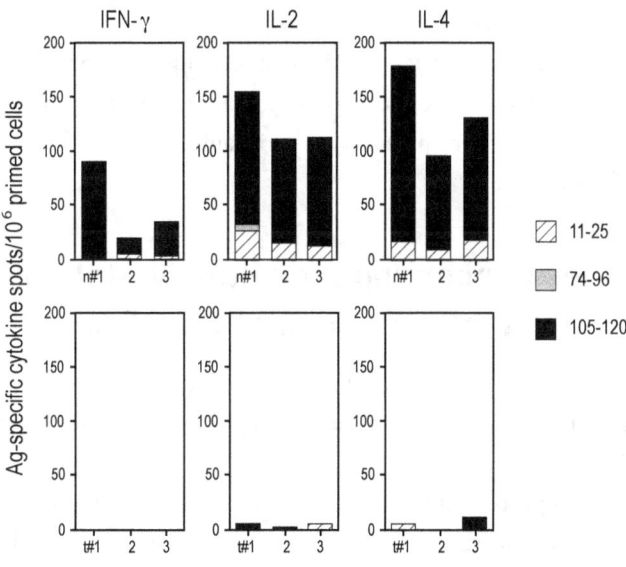

Fig. 36. Ablation of CD4 T cells specific for the major HEL peptide, $HEL_{105-120}$, abrogates the activation of CD4 T cells specific for the major peptide and for other HEL peptides normally generated upon a HEL challenge. Top three panels, responses in normal mice; bottom three panels, responses in mice tolerant of $HEL_{105-120}$. Adapted from reference 136

Observations in the literature show that CD4 T cells specific for a peptide can be ablated, i.e. rendered non-functional, by administration of large amounts of this peptide to intact mice. Given by themselves, peptides are generally incapable of activating CD4 T cells, but can usually do so if given in CFA. This was one of several facts that Janeway so colorfully referred to as the "immunologists' dirty little secret". Prior administration of the peptide without adjuvant dramatically ablates the CD4

T cell response normally observed on a subsequent challenge with the peptide given in CFA.

Nathan gave BALB/c mice a course of the major HEL peptide, $HEL_{105-120}$, in a manner known to ablate the CD4 T cells specific for this peptide, and then challenged these and normal mice with HEL. The generation of CD4 T cells specific for $HEL_{105-120}$ was almost totally ablated, as anticipated. In addition, the CD4 T cell response to the other HEL peptides[135,136] were also ablated, see Figure 36. We inferred that the CD4 T cells specific for the major peptide are necessary to obtain the efficient activation of CD4 T cells specific for other HEL peptides upon HEL challenge. This inference was further supported by the fact that peptide-exposed and normal mice made similar anti-HEL and anti-OVA responses on challenge with the HEL-OVA conjugate, except for the lack of response to the major peptide in the mice pre-exposed to this peptide. Thus, in the absence of CD4 T cells specific for the major peptide, OVA-specific CD4 T cells can facilitate the activation of CD4 T cells specific for minor HEL peptides.

This demonstration of the facilitation of the antigen-dependent activation of CD4 T cells, by other CD4 T cells, provides further support for the explanation of the phenomenon of epitope-spreading, seen in autoimmunity,[104, 105] as previously outlined in Chapter 5.

Reservations concerning the "Steady State Model" for the inactivation of peripheral CD4 T cells

It is worthwhile to consider more carefully what most immunologists currently believe are the critical circumstances determining whether antigen activates or inactivates CD4 T cells.

Observations appear to show that the uptake of antigen by immature dendritic cells (DC) can result in the subsequent inactivation of CD4 T cells specific for this antigen.[137, 138] According to the PAMP/Danger Models, this inactivation of CD4 T cells is because the DC present the antigen without sufficient expression of costimulatory molecules. In contrast, antigen-exposed DC, activated by PAMPs, appear capable of activating CD4 T cells. These observations led Steinemann to envisage that in the *steady state*, which occurs in the absence of PAMPS or danger, DC will present antigen in a manner that inactivates CD4 T cells. In the presence of PAMPS or danger, however, the dendritic cells mature, leading to greater expression of costimulatory molecules and their abilty to activate naïve CD4 T. The immune system is thus envisaged to respond to infectious entities, and perhaps to non-infectious antigens in the presence of danger, whilst non-infectious antigens, in the absence of danger, will inactivate their corresponding CD4 T cells. This view as to how antigen interacts differently with naïve T cells to result in their inactivation and activation is referred to as "the steady state model" of peripheral tolerance at the level of CD4 T cells.[137, 138] It is envisaged that these processes guarantee to a sufficient degree the inactivation of CD4 T cells specific for peripheral self antigens, so that the requisite degree of peripheral tolerance is achieved. I shall for convenience refer to this model as the ***Steady State Model of Peripheral CD4 T Cell Tolerance***.

The characterization of how immature dendritic cells take in exogenous antigen and, when mature, migrate to the draining lymph nodes, where they can activate their corresponding CD4 T cells, provides an attractive picture. This scenario accounts for how information about self and foreign antigens present in the periphery, at the site where the immature dendritic cell matures, is brought to the draining lymph node. In addition, it seems likely that the uptake of antigen by immature dendritic cells can lead to CD4 T cell inactivation and thereby contributes to the peripheral tolerance of CD4 T cells. The Steady State Model of Peripheral CD4 T Cell Tolerance accounts for these observations. It is believed by most that this model provides a comprehensive description for what is critically important in determining whether antigen activates/inactivates CD4 T cells. Although I find this steady state picture an attractive and partial description of the events that take place, I find it difficult to accept this model as providing a *full and adequate* description of the critical events controlling the activation/inactivation of CD4 T cells for two reasons.

First, there is a factual consideration. Non-infectious, vertebrate antigens, such as sterile xenogeneic red blood cells, can be potently immunogenic when administered under non-dangerous conditions. It would be expected, according to my understanding of the Steady State Model, that such antigens would be taken up by immature dendritic cells and, in the absence of danger or PAMPS, would inevitably inactivate their corresponding CD4 T cells. On the basis of this consideration, I suggest that more must be going on besides that explicitly envisaged in the steady state model, a point to which I shall shortly return.

Second, there is a conceptual consideration. The idea that whether the activation/inactivation of CD4 T cells occurs is at the mercy of microbes, pathogens in particular through their expression of PAMPS, is a mechanism that would likely lead to immunological anarchy. Anarchy would only be avoided if APC activation by PAMPs was limited to those APC presenting peptides only derived from the pathogen. How such a limitation might be achieved seems problematical and, without such a limitation, the infection by microbes would inevitably lead to the activation of anti-self CD4 T cells specific for peripheral self antigens. If danger were the critical factor, it would similarly allow the activation of CD4 T cells specific for peripheral self and for foreign antigens, if activation occurred in the locale where danger existed. These models do not minimize interference. They are inconsistent with the Historical Postulate.

For these reasons, I still feel that mechanisms of CD4 T cell inactivation/activation, consistent with the Historical Postulate, are the only way of guaranteeing self-nonself discrimination in a reliable way, which I personally still feel is the critical issue.[139] The Two Step, Two Signal Model also accounts for the fact that depletion of B cells in humans appears to be effective as a treatment of autoimmunity, including cell-mediated autoimmunity, as reviewed in reference 140. Recent studies by David Kroeger, when a graduate student, support the idea that the CD4 T cell collaboration, which can facilitate the activation of p1-specific CD4 T cells by p2-specific CD4 T cells, requires the two peptides, p1 and p2, to be presented by the same APC. Moreover, B cells, but not dendritic cells, appear to be able to mediate such cooperation.[141] I would draw the attention of those readers who are interested in a detailed discussion of contemporary issues, regarding the activation and inactivation of CD4 T cells, to a Discussion Forum in the Scandinavian Journal of Immunology.[139, 142, 143]

Further studies on immune class regulation

The generation of primary cell-mediated and antibody responses in vitro

Ian Ramshaw and I left Canberra at roughly the same time in 1977, and both moved to Canada—Ian to Queen's University as a postdoctoral fellow, and I as a faculty member of the University of Alberta. Here I joined a small group that soon included Tim Mosmann, who later became famous with Bob Coffman for their studies leading to the characterization of Th1 and Th2 clones.

Ian and I agreed that the next critical step in both our investigations would be to establish in vitro systems in which an antigen could induce cell-mediated immunity, in the form of DTH and, under different circumstances, the production of antibody. We envisaged that the development of such in vitro systems would allow one to establish what conditions favored the induction of one class of immunity over another, and why. Bob Mishell and Dick Dutton had earlier established cultures of mouse spleen that supported the antigen-dependent generation of antibody responses.[144] It was natural to employ this in vitro system in trying to realize our aims.

Ian and I again decided to do parallel studies in similar but different systems, and to keep in touch. Ian was a much more experienced experimentalist than I, and he took an embarrassingly long lead. He refrained from submitting his work[145] for publication for several months until I had caught up. This kind of collegial behaviour is almost unimaginable today, and I am still grateful to him. We published our observations at roughly the same time. I believe this was the first time that the in vitro generation of primary, antigen-specific cell-mediated responses was reported to an antigen other than to MHC antigens. There were interesting differences in the systems developed by Ian and myself. He concentrated on how the dose of antigen affected the class of immunity induced.[145] Remarkably, his findings paralleled Chris Parish's in vivo findings made in rats,[62] see Figure 21. They showed that medium doses of antigen generated IgM antibody responses and little DTH, whereas much higher and lower antigen doses generated little antibody but T cells that mediate DTH. Ian seemed to have generated in vitro analogues of the in vivo phenomena of low-zone and high-zone cell-mediated immune deviation.

I am going to discuss my experiments in some detail. I do so because I believe they provide a foundation for discussing simple strategies of immunological intervention in several situations of medical interest.

In contrast to Ian's studies, in which he varied the dose of antigen, I concentrated more on how the number of spleen cells put into a dish affected the generation of antibody and cell-mediated responses, when the amount of antigen is held constant.[146]

A methodology for showing that cellular interactions are required, or facilitate, immune responses

To understand the experimental approach we undertook, it is important to appreciate the basis of some of our quantitative approaches. These approaches allowed us to analyze whether or not, for example, the activation of a B cell, resulting in the generation of antibody-producing cells, requires cellular cooperation.

Consider the case where we take a number of spleen cells, say five million, and put these spleen cells, together with an optimal amount of the antigen *sheep red blood cells (SRBC)*, into a well of a tissue culture plate, resulting in the generation of 400 anti-SRBC IgM antibody producing cells, detectable five days after initiation of culture. How can we determine whether the activation of the B cell, leading to the generation of antibody producing cells, requires merely the binding of antigen to the receptors of the B cell, or requires or is facilitated by cellular collaboration?

Consider what we would expect to happen in the following context if the mere binding of antigen to the B cell is sufficient to activate it to generate antibody-producing cells. We take five million spleen cells and dispense them into one well with an optimal amount of antigen, and we take another five million and dispense them equally into ten similar wells, each well receiving half a million spleen cells, as well as an optimal amount of antigen. When we harvest the five million spleen cells cultured under these two conditions after five days of culture, we would expect, if the mere binding of antigen to a B cell is sufficient to activate it, that the same number of antibody producing cells would have been generated when the five million spleen cells are cultured under these two conditions. A B cell plated in relatively crowded well would not behave differently on interacting with antigen than a B cell in a less crowded well.

In practice, when we harvest the one well set up with five million spleen cells after five days of culture, and we also harvest and pool the ten wells each set up with half a million spleen cells, and we assess the total number of antibody-producing cells generated, we find that the two sets of harvested cells contain different numbers of antibody producing cells. We find that the cells harvested from one well set up with five million spleen cells contain 400 antibody producing cells, whereas the cells, harvested and pooled from ten wells, contain only ten antibody producing cells. We can conclude from this that the anti-SRBC B cells present in five million spleen cells are activated to produce antibody-producing cells forty times more efficiently when all the five million cells are in one well rather than when they are equally dispersed in ten wells. Why? It must be because the activation of a B cell, to produce antibody-producing cells, depends upon whether there are other cells, close by, that are required for or facilitate its activation. Such required/facilitating cells are much closer at hand when there are five million cells per well than when there are half a million cells per well. These observations exemplify the essence of the experiments we did, in many variations, to test the Threshold Hypothesis.

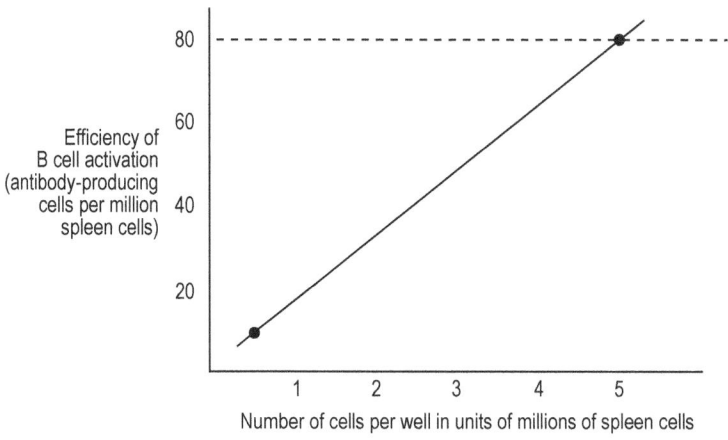

Fig. 37. Illustration of a fictional experiment to illustrate how cell cooperation in the activation of B cells can be demonstrated.

As already noted, the activation of anti-SRBC B cells is forty fold more efficient when the spleen cells are cultured at five million than when cultured at half a million spleen cells per well. We will always plot, in the figures that represent our actual observations, a parameter that represents the efficiency of activation of specific lymphocytes. This efficiency is estimated by quantitating the number of effector cells generated from a given number of cells cultured. Figure 37 represents the observations of the hypothetical experiment just described. Five million spleen cells gave rise to the 400 antibody producing cells and so resulted in a level of production of eighty (400/5) antibody producing cells per million spleen cells cultured, when all were put into one well. In contrast, when there are half a million spleen cells per well, there is a level of production of two (10/5) antibody producing cells per million spleen cells cultured. It is expected, if B cell activation does not require cell collaboration, that the production of antibody producing cells per million cells cultured will be independent of the density at which the spleen cells are cultured. This is indicated in Figure 37 by the dotted, horizontal line. Note that an upward slope indicates that cellular cooperation is required, or that such cooperation facilitates, the generation of the cells whose numbers are assessed . In practice, we culture our spleen cells at more than two densities, as we shall see, so the shape of the efficiency curve, shown in Figure 37 as a straight line, is better defined.

Different in vitro conditions support the optimal generation of cell-mediated and humoral responses

We found that optimal amounts of antigen do not induce low densities of spleen cells to generate any immune response that we assessed; the same amount of antigen induces medium densities to generate DTH-mediating cells, and higher densities of spleen cells support the sustained production of antibody and only the feeble generation of cells mediating DTH,[146] see Figure 38. These observations show that both the generation of DTH-mediating cells and of antibody producing cells requires some form of cellular cooperation. We exploited this system in diverse ways.

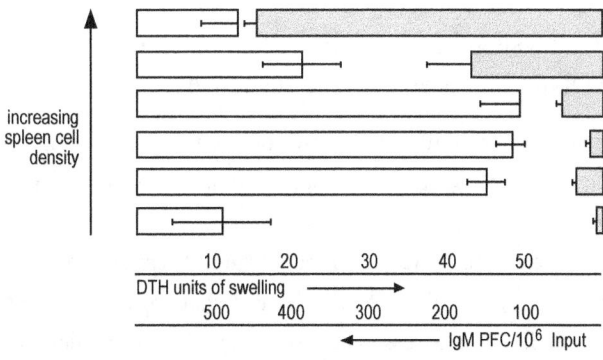

Fig. 38. A constant amount of antigen stimulates low density cultures of unprimed spleen cells to generate poor responses, a medium density to produce DTH-mediating cells, and a high density the generation of IgM-producing cells but not of DTH-mediating cells. Adapted from reference 146.

This system allowed my graduate student, Jane Tucker, to demonstrate that the generation of DTH-mediating cells requires, or is facilitated by, CD4 T cell cooperation, mediated by the recognition of linked epitopes. Thus, if Jane gave a few T cells, primed in vivo to *fowl gamma globulin (FGG)*, to low density cultures of spleen cells with the antigen FGG-SRBC, the cultures could support the generation of SRBC-specific DTH-mediating T cells. Such SRBC DTH-mediating T cells were not generated in the absence of the FGG-primed cells, nor in their presence unless the FGG was *present and linked* to the SRBC. We interpreted these observations as evidence that the generation of DTH mediating T cells requires T cell cooperation mediated by the recognition of linked epitopes.[97]

In vitro experiments allow a powerful analysis of the cellular and molecular interactions required to generate cell-mediated and antibody responses. However, it is sometimes unclear whether what happens *in vitro* truly reflects what happens *in vivo*. Therefore, in the 1980s I developed a system that allowed an experimental analysis, without prejudice, of what cellular interactions determine whether antigen induces *in vivo* a predominant cell-mediated or humoral response. The most salient observations made in this *in vivo* system parallel those made in the *in vitro* system,[148] supporting the belief that the *in vitro* system faithfully reflects the most critical *in vivo* events. I will give one or two examples of these different studies that extended over more than thirty years. I believe they provide strong evidence for the Threshold Hypothesis. I do so not only because the nature of the decision criterion, controlling whether antigen induces cell-mediated or humoral immunity, is a fundamental question, but because I believe the answer to this question is most likely relevant to the realization of several important medical goals, as I shall expand upon later. These goals include reversing AIDS shortly after seroconversion, and developing effective therapy against some cancers.

Experimental analysis of the "decision criterion" determining Th1/Th2 phenotype of in vivo immune responses

Most of the *in vivo* experiments I will describe are relatively recent, done fifteen to twenty years ago, and constitute the research that Nahed Ismail presented in her PhD thesis.

Nahed lethally irradiated mice and reconstituted them the same day, day 0, with different numbers of syngeneic spleen cells, and challenged all mice with the same amount of antigen. She then assessed at various days how the efficiency of the generation of Th1 and Th2 cells, and of IgM and IgG antibody producing cells, depended on the number of spleen cells employed for reconstitution. Figure 39 represents our findings, as assessed on day 6, by which time substantial responses are generated.[148] Three points struck us. First, the efficiency of the generation of IFN-γ-producing Th1 cells did not vary significantly in the irradiated mice reconstituted with different numbers of spleen cells, over the range of spleen cells used for reconstitution, from about 2 to 80 million. Second, and in contrast, the efficiency in the generation of IL-4-producing Th2 cells increased dramatically as the number of spleen cells employed for reconstitution was increased. Third, the generation of IgG antibody responses was cell cooperation dependent and this dependency reflected, in its general tendency, the generation of IL-4-producing T cells.

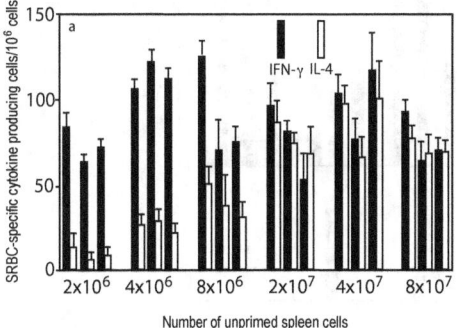 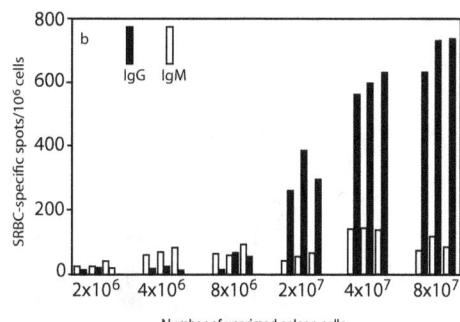

Fig. 39. The dependency of the efficiency in the generation of IFN-γ- and IL-4-producing cells, IgM and IgG antibody-producing cells on the number of unprimed spleen cells employed to reconstitute lethally irradiated mice. Responses of individual mice are shown three per group. Adapted from reference 148.

Clearly, more unprimed spleen cells are required to efficiently generate Th2 than Th1 cells. Can a particular cell type be identified that is required in greater numbers to generate Th2 than Th1 cells? We reconstituted irradiated mice with a mixture of unfractionated and fractionated unprimed spleen cells and immunized them, as indicated in Figure 40. The particular experiment shown identifies the cell type, needed in greater numbers to generate Th2 than Th1 cells, as a CD4 T cell. These experiments demonstrate that more CD4 T cells are required to generate Th2 than Th1 cells, when the amount of antigen is held constant,[148] a prediction of the Threshold Hypothesis.

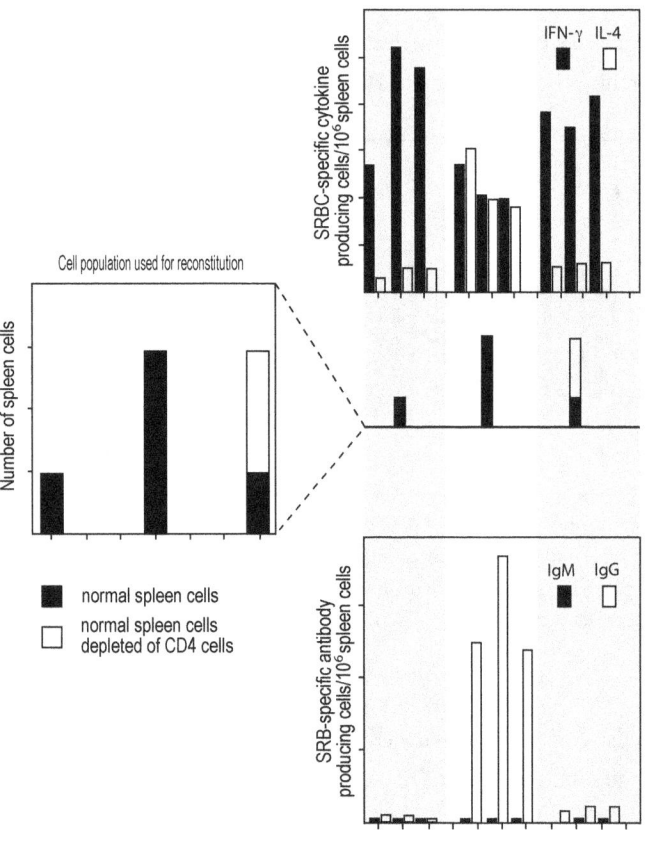

Fig. 40. Evidence that the generation of SRBC-specific IL-4-producing Th2 cells requires more CD4 T cells than does the generation of IFN-γ-producing, Th1 cells. The responses of individual mice, three per group, are shown. Adapted from reference 148.

Further *in vivo* experiments test this interpretation even more directly and provide some additional insight into the mechanism of CD4 T cell cooperation, see Figure 41. Lethally irradiated mice were all reconstituted with 2×10^6 million spleen cells and immunized with various amounts of different antigens, and sometimes with additional lymphoid cells from the DO11.10 mouse transgenic for a T cell receptor that recognizes a peptide derived from ovalbumin (OVA) in the context of host MHC class II molecules. The mouse is thus a highly enriched source of OVA-specific CD4 T cells. This experimental system allowed us to systematically test predictions highly specific to the Threshold Hypothesis. We illustrate these studies with the observations recorded in Figure 41. It will be easier to follow this experiment if we outline the gist of the ideas it was designed to test.

The Threshold Hypothesis explains how both the dose of antigen, and the number of CD4 T cells, determine the Th1/Th2 phenotype of the Th cells generated. Moreover, the Th1/Th2 phenotype depends upon these two variables in an interdependent fashion. Recall the situation where a moderate amount of antigen and moderate number of CD4 T cells results in robust CD4 T cell cooperation, leading to the generation of Th2 cells. We can reduce the strength of CD4 T cell cooperation by either reducing the amount of antigen, or by reducing the number of CD4 T cells. If we reduce either sufficiently, to what we refer to as a low level, the CD4 T cell cooperation will be tenuous and Th1 cells will be generated. Consider the situation where the amount of antigen is low and where Th1 cells are generated. We can increase the level of CD4 T cell cooperation by increasing the number of CD4 T cells, so that now CD4 T cell cooperation is again robust, and Th2 are again generated. The observations in Figure 41 test the validity of these expectations.

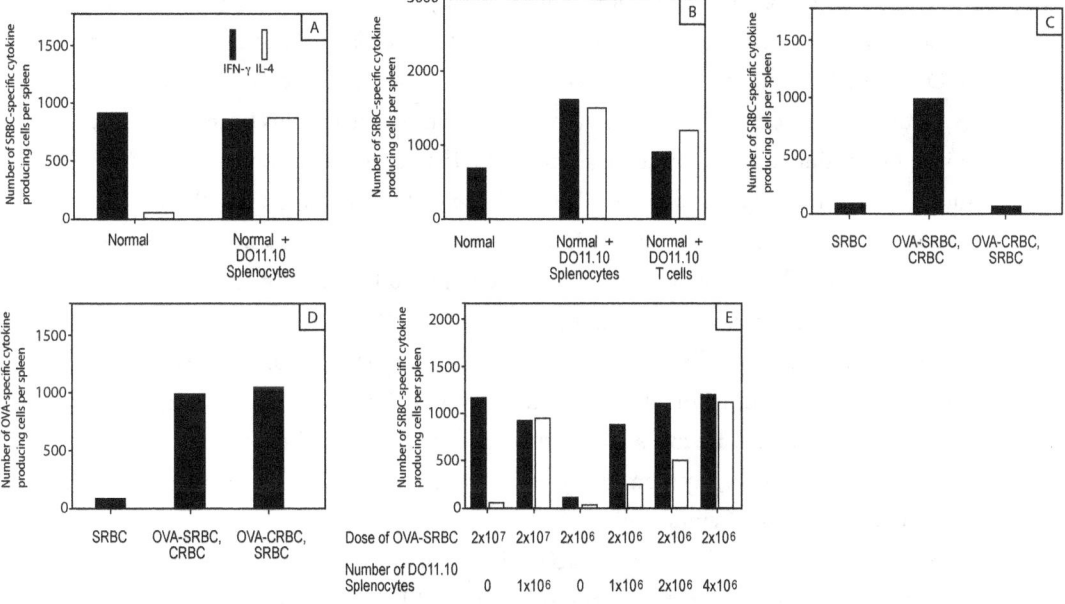

Fig. 41. The number of responding CD4 T cells and the dose of antigen jointly affect the generation of Th1 versus Th2 cells in vivo. **Panels A-D** Lethally irradiatd BALB/c mice were reconstituted with either 2×10^7 syngeneic, unprimed spleen cells alone or with an additional 1×10^6 spleen cells, or CD4 T cells purified from 1×10^6 spleen cells from DO11.10 transgenic mice. The mice whose response in recorded in panels A and B were immunized with 2×10^7 OVA-SRBC. The mice whose response in recorded in panels C and D were immunized with either 2×10^6 SRBC, 2×10^6 OVA-SRBC and 2×10^6 CRBC, or 2×10^6 SRBC and 2×10^6 OVA-CRBC, as indicated, and in addition 10^6 DO11.10 transgenic spleen cells. **Panel E** Mice were irradiated and reconstituted with unprimed spleen cells and given additional number of DO11.10 spleen cells as shown, and immunized with the indicated number of OVA-SRBC.

We can see from the observations recorded in panel A and panel B that the additional presence of OVA-specific CD4 T cells, in the form of spleen cells or puri-fied splenic T cells from the transgenic mice, allow the antigen OVA-SRBC, which would otherwise generate a predominant SRBC-specific Th1 response, to generate substantial numbers of IL-4-producing, SRBC-specific Th2 cells. In panel C, the irradiated mice received a ten-fold lower dose of various antigens and, in addition to the standard $2x10^7$ normal spleen cells, they all received 10^6 transgenic spleen cells as a source of OVA-specific CD4 T cells. Mice challenged with OVA-SRBC and CRBC produce significant numbers of IFN-γ-producing SRBC-specific Th1 cells, but not when given either SRBC or SRBC and OVA-CRBC. Thus OVA-specific transgenic CD4 T cells can only help the generation of SRBC-specific IFN-γ producing CD4 T cells when OVA is present and linked to the SRBC. Panel D shows that both OVA-CRBC and OVA-SRBC generated similar and significant numbers of OVA-specific IFN-γ-producing cells under these circumstances, as expected.

The observations in panel E show that the ability of OVA-specific CD4 T cells to facilitate the generation of SRBC-specific Th1 and Th2 cells depends on the number of OVA-SRBC employed in the challenge, in the manner expected on the Threshold Hypothesis. For example, the same number of a million spleen cells from transgenic mice will modulate a predominant anti-SRBC Th1 response to a mixed Th1/Th2 mode, when the antigen challenge is $2x10^7$ OVA-SRBC, but will support the generation of a Th1 response when the antigen challenge is ten fold less, i.e. $2x10^6$ OVA-SRBC. Additional transgenic CD4 T cells are required to generate substantial Th2 cells with this lower amount of antigen, as anticipated.[149]

These and related observations test critical predictions of the Threshold Hypothesis. However, it might be objected that lethally irradiated mice, reconstituted with spleen cells, are in a state of stress under which lymphocytes may behave in some respects differently from those in intact mice. We therefore also tested the Threshold Hypothesis in intact mice.

Some years earlier, I had decided to explore whether such an *in vivo* study was feasible, employing mice that were transgenic for HEL, and whose immune system therefore regarded HEL as a self antigen. These transgenic mice had been developed by Chris Goodnow and Tony Basten. I went to Sydney, Australia, for a couple of months, to carry out the first experiments with Tony. The preliminary results seemed promising. Nahed undertook a formal study based upon these preliminary findings.

Nahed gave some mice a low dose of SRBC that generates a predominant Th1 response, and other mice a higher dose that generates a predominant Th2 response, as assessed by the appearance of SRBC-specific cytokine-producing CD4 T cells six days after immunization, see Figure 42. She then assessed the effect of coupling a foreign antigen to the SRBC and giving a low dose that results in a Th1 anti-SRBC response with uncoupled RBC. The Pearson and Raffel generalization implied, in a modern context, that making an antigen more foreign will increase the Th2-component of the immune response to the antigen. This happened when we coupled either ovalbumin (OVA), or hen egg lysozyme (HEL), to the SRBC, and immunized normal mice. We next investigated why coupling a foreign protein to the SRBC facilitated the Th2 component of the anti-SRBC response. We were aware that the mere procedure of coupling the protein antigens to SRBC might change the physical and immunogenic properties of the coupled cells; in addition, we were also aware that, by coupling these foreign antigens to the SRBC, we were making them more foreign, i.e, recognized by more CD4 T cells.

To explore whether the latter possibility contributed to the nature of the response to coupled SRBC, we employed HEL-transgenic mice that are tolerant of HEL at the level of the T cell population. We found that similar anti-SRBC responses are generated when normal mice and HEL-transgenic mice are immunized with various forms of the antigen, with one exception. The response to SRBC on challenge with HEL-SRBC, see Figure 42, was different in normal and in HEL-transgenic mice: the Th2 component was enhanced, at the expense of the Th1 component, to a greater extent in normal than in HEL-transgenic mice. Thus these observations appear to show that HEL-specific T cells facilitate the Th2 component of the anti-SRBC response when normal mice are challenged with HEL-SRBC, but that such facilitation does not occur to the same extent in HEL-transgenic mice, that have a deficiency of HEL-specific T cells. These observations show that increasing the T cells specific for an antigen can increase the Th2 component of the in vivo response,[150] thus confirming our explanation of the Pearson/Raffel generalization, as incorporated in the Threshold Hypothesis.

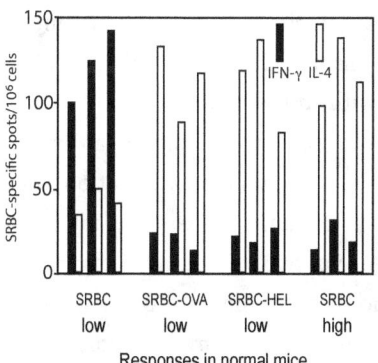 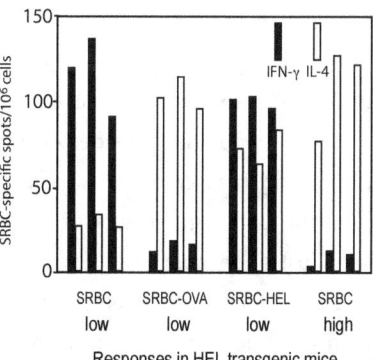

Fig. 42. HEL-specific T cells help increase the Th2 component of the anti-SRBC response when intact mice are challenged with HEL-SRBC. Responses of individual mice, three per group, are shown.

Coherence of immune responses and immune responses to real pathogens

It is clear that the Th1/Th2 phenotype of an immune response generated against a pathogen is often of critical importance in containing the outcome of the infection. We also understand to a certain extent why this is so at a mechanistic level. *Mycobacterium tuberculosis* infects and replicates inside macrophages. Mycobacterium-specific Th1 cells can deliver IFN-γ to a macrophage infected by *M tuberculosis*, and the IFN-γ binds to surface receptors of the macrophage to activate metabolic pathways, inside the macrophage, that kill the mycobacteria.[151] If mycobacterium-specific IL-4-producing CD4 T cells are generated, they can result in the delivery of IL-4 to the infected macrophage, which in turn results in the down-regulation of these anti-microbial metabolic pathways.[152] Thus, if a mycobacterium-specific immune response is to be effective in containing *M tuberculosis,* it must not only result in the substantial generation of specific Th1 cells but avoid the generation of a substantial number of specific Th2 cells. Indeed, if one employs an assay that can detect single, antigen-specific cytokine producing cells, to examine the nature of the immune response to chemically complex pathogens, it is really remarkable how predominant Th1 and Th2 responses can be. This is not to deny that many times mixed responses occur. Note that not only does regulation tend to ensure the exclusive generation of Th1 and Th2 responses, but the example just given illustrates that at times the activity of effector Th1 and Th2 can also be counteractive.

Most of the experimental analysis, described above and successfully testing predictions of the Threshold Hypothesis, have been carried out with antigens that are both chemically rather simple and that do not replicate. Two questions might be asked concerning real, much more complex antigens, such as live bacterial and protozoan pathogens, that are the real concern of the immune system. Does the Threshold Hypothesis apply to immune responses against these microorganisms/parasites in the same way as it applies to responses to chemically relatively simple, non-replicating antigens? If the threshold mechanism holds, is it problematical to understand how this is achieved mechanistically with complex antigens?

A first point in considering immune responses against live microorganisms is important. It is known that, if a pathogen multiplies very rapidly, infection with just one organism results in a predominant Th2 response. In essence, infection with just one viable organism leads to the existence of many millions of organisms five days later. In this case, immunologists cannot control the level of the antigen load by choosing the number of organisms they infect with. Any infection invariably results within days in a high antigen load, and so a predominant Th2 response rapidly ensues. However, we can control the antigen load with intracellular parasites that grow slowly. I take examples from organisms with which we have had direct experience. Infection with relatively low and high numbers of mycobacteria and leishmania parasites respectively results in predominant Th1 and mixed Th1/Th2 or predominant Th2 responses, as we shall later describe.

There are strong reasons for believing the threshold mechanism similarly applies to immune responses to these chemically complex and slowly replicating antigens as to immune responses to more simple antigens. Thus, the Threshold Hypothesis accounts for how the Th1/Th2 phenotype of the immune response depends upon the dose of a chemically simple antigen and the time, after antigen impact upon the immune system, at which this phenotype is assessed. It turns out that the Th1/Th2 phenotype of the immune response against complex antigens, such as live and slowly growing intracellular bacteria and protozoan parasites, is similarly dependent upon these same variables of immunization. These observations provide strong, though indirect, evidence that the threshold mechanism also applies in these cases. Given this plausibility, and the more direct evidence outlined later, we consider whether there are difficulties in understanding mechanistically how the "threshold mechanism" is realized in the context of complex antigens.

Consider a *Leishmania major* parasite. There must be some components of the parasite that are relatively highly abundant, denoted as A, and others that are much less predominant, denoted as L. These different components will also generally be of different size, different in their degrees of foreignness, and so different in the number of CD4 T cells specific for them. Consider the nature of the immune response upon giving a number of dead parasites to a mouse. The amount of A will be much larger than the amount of L. There might also in general be different numbers of CD4 T cells specific for A and L. It is clear that, if A and L were non-cross-reactive antigens and physically free of one another, they would in general generate CD4 T cells of different Th1/Th2 phenotype. This envisaged sitation is just the one we described earlier, depicted in Figure 33, in which we simultaneously gave low and high numbers of two non-crossreacting foreign RBC, L and H, resulting in a Th1 anti-L and a Th2 anti-H response. In this scenario, the response to these parasites would rarely if ever be **coherent**; in other words, the cytokine profile of the large majority of the Th cells specific for these two different antigens would be different. However, we have deliberately misrepresented the situation to reveal the problem. The A and L components will not always be physically free of one another, and they may also be indirectly linked by the physical existence of A/C and L/C complexes, where C is another protein of *Leishmania major*. This is clearly a complex situation.

Let us consider a particular case. Suppose all L is associated with A as A/L complexes. As A is abundant and L is much less prevalent, there will be much more A than A/L complexes. It is again likely that A will generate CD4 T cells of a different Th1/Th2 phenotype from the phenotype of the L-specific CD4 T cells generated. However, the immune responses to these two antigens are not independent. For example, the presence of A-specific CD4 T cells will affect the Th1/Th2 phenotype of the L-specific CD4 T cells generated, as the two antigens are linked. Nevertheless, it seems clear when we consider what might be the initial Th1/Th2 phenotype of the CD4 T cells generated against the diverse proteins of *Leishmania major*, and in the context of the Threshold Hypothesis, that infection with low numbers of viable parasites may give rise predominantly to Th1 cells, and with higher numbers to a transient Th1 response that can rapidly evolve into a mixed Th1/Th2 or predominant Th2 mode. However, we might well initially anticipate that these responses are unlikely to be very coherent. My colleagues and I have been surprised by the degree of coherence we have observed. This is a point to which we shall shortly return. We shall shortly argue that the very nature of the Th1 and Th2 themselves is critical to achieving coherence!

An assessment of the role of cytokines in the determination of the Th1/Th2 phenotype of immune responses

As already outlined, another interesting and highly prevalent view is that the cytokine environment in which antigen initially interacts with naïve CD4 T cells and APC determines the Th1/Th2 phenotype of the CD4 T cells generated.

We have already seen that the presence of IL-12 and/or IFN-γ can favor the generation of Th1 cells, and of IL-4 the generation of Th2 cells. To summarize, observations employing intact mice leaves little doubt that IL-4 and IFN-γ have a pivotal role in determining the Th1/Th2 phenotype of the immune response.

These observations have led to two different views. The first, already outlined in Chapter 5, is that the ambient cytokines, when antigen first interacts with naïve CD4 T cells, determine the Th1/Th2 phenotype of the effector CD4 T cells generated. We recall, for example, that the administration of anti-IL-4 antibody, around the time of parasite infection of mice with *L major,* can prevent the generation of Th2 responses, and facilitates the generation of IFN-γ-producing Th1 cells instead. This shows that IL-4 is required to foster predominant Th2 responses. What could be the source of this IL-4? It is known that cells, other than lymphocytes, such as mast cells and other cells, can produce IL-4. Following this line of reasoning, immunologists try to explore what cells of the innate system might provide the IL4 required to support the initial generation of Th2 cells.[153-155] Similarly, investigators try to identify the sources of cytokines, in terms of cells of the innate system, that might foster the generation of Th1 responses by the production of IL-12, for example.[109] I refer to this view as the ***Cytokine Milieu Hypothesis.***

We now consider an alternative explanation for the experimentally established roles of IFN-γ and IL-4 in determining the Th1/Th2 phenotype of a response. It is convenient here to break down the envisaged process by which Th1 and Th2 cells are generated into an initial phase, involving the "threshold mechanism", that usually determines whether Th1 or Th2 cells are predominantly generated, and a later phase. According to this explanation, cytokines are not normally of critical importance at this initial phase, and indeed are likely normally not present at physiologically significant levels in the environment where naïve CD4 T cells are activated. In contrast, in the subsequent phase, effector Th cells that produce cytokines will have been generated in the first phase, according to the threshold mechanism. We suggest this later phase sharpens up the consequences of the initial phase.

We have discussed in the previous section why the processes, described by the Threshold Hypothesis in the context of complex antigens, such as slowly replicating pathogens, must often lead to a somewhat incoherent response, in the sense that rarely would we expect only Th1 or Th2 cells to be initially generated. We suggest that the second phase of the response results in a sharpening up of the implementation of the initial phase, resulting in more physiologically useful and coherent responses. The idea of sharpening up the consequences of the initial phase is supported by the nature of the activity of different cytokines made by Th1 and Th2 cells. Some of the cytokines made by Th1 and Th2 cells are either self-promoting for cells belonging to their subset, or are inhibitory for the generation of CD4 T cells of the opposing subset. For example, the IFN-γ, made by Th1 cells, preferentially inhibits the multiplication of Th2 cells,[111, 112] and the IL-4 made by Th2 cells stimulates the multiplication of Th2 cells but not of Th1 cells.[113] The IL-10 made by Th2 cells inhibits the production by Th1 cells of IFN-γ,[156] thus blocking the IFN-γ-dependent inhibition of the proliferation of Th2 cells. This alternative explanation of the role of cytokines, conveniently referred to as the *Cytokine Implementation Hypothesis*, naturally accounts for why some of these critical cytokines, IFN-γ and IL-4, are made by Th1 and Th2 cells themselves, and why the cytokines have self-promoting/self propagating properties for the cells that produce them. However, can this Cytokine Implementation Hypothesis account for the *in vivo* observations on the central role of IL-4 and IFN-γ in determining whether Th2 or Th1 responses are generated?

Consider the observations on the dramatic and required role of IL-4 in supporting Th2 responses. As we have explained above, the threshold mechanism is expected to result initially in a somewhat incoherent response. For illustrative purposes, consider a situation in which Th2 cells are initially predominantly generated. Suppose four IL4-producing Th2 cells are produced for every one IFNγ-producing Th1 cell. In the presence of neutralizing anti-IL4 antibody, IL4 will not stimulate the Th2 cells to divide. The Th1 cells will produce both IFNγ and IL2. The presence of IL2 alone would stimulate both the Th1 and Th2 cells to divide, but in the presence of IFNγ, Th1 cells are likely to preferentially divide. Thus, the effect of the anti-IL4 antibody may in time modulate a mildly predominant Th2 response into an exclusive Th1 mode!

It might seem initially that the Cytokine Milieu Theory and the Cytokine Implementation Hypothesis are somewhat similar. However, their significant differences are brought out when we consider their consequences. I illustrate this by considering again the role of IL4 in supporting Th2 responses. The Cytokine Milieu Hypothesis has led to many studies to identify the source of IL4, other than Th2 cells, believed to be required to initiate the generation of Th2 cells.[153-155] In this case, if IL4 is the critical factor, there will likely be no attempts to consider other factors/circumstances important in the generation of Th2 cells. In the case of the Cytokine Implementation Hypothesis, the nature of the mechanism controlling the initial phase in the determination of the Th1/Th2 phenotype of the response is left open, and so efforts will be made to define it.

Chris Rudulier, in his PhD studies, established an *in vitro* system in which he employed CD4 T cells from a mouse transgenic for a TcR that recognizes an OVA peptide in the context of host class II MHC molecules. This mouse was thus a ready source of OVA-specific CD4 T cells. Chris established in vitro conditions under which the pertinent peptide could activate these transgenic CD4 T cells to give rise to Th1 and/or Th2 cells. These conditions corresponded to the conditions under which polyclonal CD4 T cells could be activated *in vitro* and *in vivo*. Thus in all systems the Th1/Th2 phenotype of the activated Th cells generated depended upon both the number of CD4 T cells and amount of antigen present in a culture well or present in a mouse. Chris analyzed several mechanistic features of the CD4 T cell interactions that determine the Th1/Th2 phenotype of the activated CD4 T cells. He found, as others had, that the generation of IL-4-producing CD4 T cells is inhibited by the presence of neutralizing anti-IL-4 antibody, and that Th1 cells are generated instead. Interestingly, he could show that the pertinent IL-4 was made by the CD4 T cells themselves, as purified CD4 T cells could be activated to generate IL-4-producing Th2 cells when the non-T cells, required to allow antigen to generate activated CD4 T cells, came from an IL-4 knock out mouse. Thus, the critical IL-4 required to generate a substantial and sustained Th2 response is produced by the CD4 T cells themselves. This observation of course supports the Cytokine Implementation Hypothesis. I would also point out that the source of the IFNγ, which supports Th1 responses, may well be produced predominantly by CD8 T cells, generated together with Th1 cells. I suggest this in part because of the finding, already described, that mice, in a state of cell-mediated immune deviation, harbor CD8 TsAb cells.[77] These CD8 T cells make potent levels of IFN-γ (C Havele, pers comm.).

The Cytokine Milieu, the PAMP and the DC-type Hypotheses as explanations for the nature of the Decision Criterion controlling the Th1/Th2 phenotype of the immune response

We have just seen two ways in which the dramatic observations on the role of IL-4 and of IFN-γ, in controlling the Th1 and Th2 phenotype of the immune response against *L major*, can be accommodated, namely by the very different Cytokine Milieu Hypothesis and the Cytokine Implementation Hypothesis. In general terms, which is correct? First, there is no *a priori* reason to believe that these alternatives are necessarily incompatible in the general case. My own view is that something akin to the Cytokine Milieu Hypothesis is pertinent in some cases, for instance in understanding some aspects of how immunity against gut flora is generated. In this case, APC associated with the gut are different from APC present in, for example, the spleen. These gut-associated APC tend to favor, through their production of retinoic acid and possibly other characteristics, the generation of CD4 T cells that, upon chronic stimulation, facilitate the generation of "regulatory T cells"[123], and so help in IgA and IgG$_4$ antibody responses. Nevertheless, in the instance of trying to understand the role of IFN-γ and IL-4 in determining the Th1/Th2 phenotype of the CD4 T cells, I favor the account provided by the Cytokine Implementation Hypothesis for reasons beyond those outlined above. To explain why, it is useful to note some further observations made in the mouse model of cutaneous leishmaniasis.

We recall that different strains of mice, given a standard challenge of a million *L major* parasites, are denoted respectively as resistant or susceptible, depending on whether the parasite is contained at the site of infection, or increases in number and metastasizes to distal sites. Resistance is associated with a sustained Th1 response and susceptibility with a transient Th1 response that evolves into a predominant Th2 mode.

We showed that susceptible mice, infected with a hundred or a thousand parasites rather than the standard challenge of a million parasites, generate a prolonged, sustained, and stable Th1 response. It was also found that partial depletion of CD4 T cells in susceptible mice, around the time of challenge with a million parasites by administering a monoclonal antibody specific for CD4, render the mice resistant, which is again associated with a predominant and stable Th1 response. These findings parallel our work with simple antigens, showing that more CD4 T cells are required to generate Th2 than Th1 cells, all other variables being kept constant. This last observation can be predicted on the basis of the Threshold Hypothesis. The dependence of the Th1/Th2 phenotype of the immune response generated in susceptible mice, on the number of parasites used for infection, on the time after infection at which this phenotype is assessed, and on the number of CD4 T cells present, are all accounted for by the threshold mechanism.

None of these three types of observation are easily reconcilable with the Cyotokine Milieu Hypothesis, the PAMP Theory, or the DC-type Hypothesis. In particular, none of these alternatives can explain how the number of CD4 T cells present determines the Th1/Th2 phenotype of the ensuing response in the observed manner, when all other variables are kept constant. For illustrative purposes, consider the situation where the number of parasites employed for infection, and the time after infection at which the Th1/Th2 phenotype is assessed, is varied in the context of the PAMP Theory. The evolution of the immune response from a predominant Th1 to a mixed Th1/Th2 and then to a Th2 mode is incomprehensible, as the nature of the PAMPs does not change with time. It is also difficult to explain the dependence of the Th1/Th2 phenotype on the number of parasites within the context of this framework. In addition, we have seen a similar dependence of this phenotype on antigen dose and time after immunization for vertebrate antigens such as SRBC.[71, 148] These antigens, from a vertebrate source, are not expected to contain PAMPs, and so the nature of their non-existent PAMPs cannot explain the pertinent observations made with these antigens.[146] It seems that any explanation for these dependencies of the Th1/Th2 phenotype of immune responses, in the context of these non-PAMP containing antigen, can similarly explain the same dependencies for PAMP-containing antigens. It thus seems likely that the nature of the PAMPs are generally not critical in determining the Th1/Th2 phenotype of most immune responses. I suggest similar arguments can be made against the DC-type Hypothesis.[116] In particular, this hypothesis cannot explain why the number of CD4 T cells determines the Th1/Th2 phenotype of the response, all other variables being held constant.

These considerations should not be used to deny the role of PAMPs or different types of DC in other important contexts. For example, it appears that gut colonization by certain bacteria is required, or facilitates, the generation of responses mediated by Th17 cells.[157] This observation supports the idea that the PAMPs of these bacteria can be critical in priming Th17 cells. The APC present in gut lymphoid tissue are different from other APC from non-mucosal lymphoid tissue, and favor the generation of T_{reg} cells,[123] as already discussed in Chapter 5.

The Nobel Prize for Physiology or Medicine, 2011

The section was inserted and first drafted during the week the 2011 Nobel Prize for Physiology and Medicine was announced. Bruce Beutler, Jules Hoffmann, and Ralph Steinman have made major contributions to the field. Lemaitre and Hoffmann discovered Toll receptors in the common fruitfly, which are critical for defense. Fruitflies without these receptors are highly susceptible to bacteria and fungi. These insects do not have an immune system in the sense that vertebrates do, and these Toll receptors can be regarded as part of innate defense.

Bruce Beutler discovered similar receptors in mice. These receptors were christened *Toll-like-receptors*, or *TLR* molecules. A series of different types of TLR molecules are found on or in phagocytic cells that act as APC. Ralf Steinman discovered a most significant cell, the dendritic cell, which acts as an APC. We have seen that immature dendritic cells take in foreign and self matter from the sites where they reside, and with time these dendritic cells mature. The mature dendritic cells are no longer phagocytic, express more class II MHC molecules and costimulatory molecules on their surface than the immature dendritic cells from which they are derived, and migrate to the nearest lymph node or to the spleen. They thus provide the lymphocytes in these secondary lymphoid organs with information as to what is going on at the site of their initial residence, when immature.

For example, if bacteria are present at this site, the immature dendritic cells will engulf the bacteria, mature, and then migrate to the draining lymph node and present bacterial peptides to the resident T cells. As we have seen, the process of maturation of DC can be greatly facilitated when bacterial components interact with the TLR of the immature DC. This is an example of a way by which the recognition of PAMPs by cellular PRR can facilitate the generation of substantial and more rapid immune responses. The consequences of the presence of PAMPS on immune responses make eminent sense. It is natural that these signals of alarm are exploited by the immune systems to increase the size and the tempo of the immune response. It is easy to imagine the evolutionary and selective forces that result in more rapid and more intense immunity when pathogens are detected inside the body.

However, most immunologists go further than this. Indeed, I and many other immunologists thought that Janeway would most likely have replaced one or another of the awardees had he been alive. Most immunologists see the findings just outlined as a validation of Janeway and Matzinger's proposal that the presence of such PAMP/Danger signals not only determines whether CD4 T cells are activated, but that in the absence of such PAMP /Danger signals the antigen will inactivate the CD4 T cells. It is also often suggested that the Th1/Th2 phenotype of the ensuing immune response is determined by the PAMPs present. Thus, PAMPs/Danger signals are envisaged not only to affect the generation of immune responses, but to *instruct* the immune system as to whether or not an immune response is to be launched and, if so, what type of response should be generated, i.e. class of immunity expressed. I quote from the first report on the web today concerning the awarding of this Nobel Prize:

> After the initial detection of foreign invaders by the Toll or Toll-like receptors, dendritic cells process the danger signals from these receptors and then *instruct* (my emphasis) other immune cells called T cells to fight infection and develop a memory of the invading pathogen, says Alexander G. Betz, who studies adaptive immunity at the MRC Laboratory of Molecular Biology, Cambridge, England.

A major focus of the view developed here is that circumstances other than the presence of PAMP/Danger signals are usually central in determining whether CD4 T cells are activated, and in determining the Th1/Th2 phenotype of the effector CD4 T cells generated. I would just like to develop one thought here that I believe provides some insight into why we find ourselves with the view that is contemporarily prevalent.

Historically, the development of ideas on the signal importance of PAMPs/Danger followed Janeway's recognition that it was necessary to use microbial adjuvants to readily obtain immune responses against antigens employed in many classical immunological studies. This recognition was the inspiration for Janeway's famous reference to the "immunologists' dirty little secret." Can we understand why this came about, without following Janeway's path? I believe we can.

Immunologists in the 1960s and 1970s were interested in what circumstances resulted in the ability of antigen to generate an antibody response, and under what other circumstances the administration of the same antigen could generate an unresponsive state. Some antigens are very immunogenic, and can readily induce vigorous antibody responses. One example would be sheep red blood cells (SRBC) when administered to mice. In practice, it is difficult to find conditions under which the administration of very immunogenic antigens can induce an unresponsive state. Thus studies with such antigens are generally not very enlightening in providing information on the different circumstances under which antigen activates and inactivates lymphocytes. Studies with such antigens were unhelpful in providing clues as to what was required to obtain an antibody response, as such responses are so readily attainable.

In contrast, some other antigens were found to be either non-immunogenic, or very poorly immunogenic, depending upon the circumstances of their administration. It was possible to explore, with such antigens, what conditions made the antigens more immunogenic and to thereby obtain clues as to what was required to generate immune responses. In addition, it was also possible with some of these antigens to define conditions under which they generated unresponsive states. I recall four examples extensively discussed in previous sections. Poly-L-lysine (PLL) is a hapten in some guinea pigs, as is the B_4 porcine lactic dehydrogenase (LDH) enzyme in some rabbits. In both cases, coupling the haptenic PLL and the LDH enzyme to an immunogenic carrier resulted in an antibody response to the macro-molecular hapten. Such findings led in time to the recognition that a B cell/T helper cell interaction is required to generate most antibody responses. Other studies employed bovine serum albumin (BSA) in rabbits and guinea pigs. Deaggregated BSA was non-immunogenic in guinea pigs, and could be used to generate an unresponsive state. The importance of the experimental system in which BSA was given to neonatal rabbits to generate unresponsive states has been extensively discussed, see Figure 12 of Chapter 3

All these antigens, besides PLL to which we shall return, have one property in common. These foreign antigens are derived from other vertebrate species that have homologous self-proteins. In a sense, and for this reason, they are not very foreign, as many of the lymphocytes generated with receptors able to recognize the foreign antigen will also be able to recognize the homologous self antigen, and so be eliminated by self-tolerance mechanisms. Moreover, it seems that the closer the antigen donor is, in the evolutionary sense, to the animal immunized, the less foreign it is expected to be and this correlates with it being less immunogenic, as expected. In addition, the lower the molecular weight of the homologous protein, the less immunogenic it tends to be. This is all understandable if the number of helper T cells specific for an antigen depends upon the size of the antigen and its degree of foreignness, and if the initiation of the generation of immune responses to such antigens is limited by the number of specific T helper cells. This is to be expected if CD4 T cell cooperation is required to activate CD4 T cells. In these classical studies, immunologists often resorted to strong adjuvants to obtain vigorous responses to these weakly immunogenic antigens. We now know PLL is a hapten as there are few if any PLL-specific T helper cells in non-responder guinea pigs, as PLL is not presented by non-responder APC. Thus it seems these antigens, chosen so they could either generate unresponsive states or were not immunogenic under certain conditions, did so because of the relatively low number of helper T cells specific for them in the intact animal. Thus, it was helpful to use microbial adjuvants to obtain robust antibody responses.

Does the requirement for microbial adjuvant to obtain potent antibody responses to these foreign vertebrate antigens really represent a universal requirement to obtain responses to such antigens, as inferred by Janeway and elaborated upon by Matzinger? More complex vertebrate antigens, containing many foreign and linked antigens, for which there are more helper T cells, such as sterile SRBC in mice, are highly immunogenic without adjuvant. It is difficult to see how the PAMP/Danger Models can accommodate such findings in their original and strict formulations.

Speculations on the regulation and function of states of anti-inflammatory immunity

The Hygiene Hypothesis and its variants

There has been a great deal of interest in the immunological and broader community concerning the basis of the increasing prevalence of allergic-dependent asthma in industrialized countries. This is the form of asthma in which attacks are triggered by inhalation of an allergen against which the individual has substantial IgE antibody, resulting in an acute inflammatory response in the lung. This increased prevalence of asthma causes considerable suffering. The Hygiene Hypothesis, as originally formulated, postulated that hygienic practices in industrialized societies lead to less exposure to a diversity of pathogenic and non-pathogenic organisms, particularly in the young. It was proposed that such exposure in non-industrialized countries led in time to Th1-imprints and biased immune responses away from those with predominant Th2 modes that lead to allergic-dependent asthma.[158]

Although some findings appeared to support the Hygiene Hypothesis, one further consideration and one further observation led to questions of its plausibility, as originally formulated. First, the proposition that greater exposure to a diversity of organisms in "non-hygienic" societies leads to predominant Th1 responses and Th1 imprints was neither well justified nor really plausible. Why would such exposure not often lead to Th2 responses and Th2 imprints? Second, there appears to be in industrialized societies not only an increase in allergic-dependent asthma, mediated by Th2 responses, but also in Th1-mediated autoimmunity. If this counter-argument is taken seriously, and if one wants to maintain the idea that hygiene is pertinent to the increase in allergy and also in Th1 autoimmunity, as seems plausible from many studies, it would appear that "non-hygienic conditions" can lead to "immunological" states unfavourable to both Th1 and Th2 responses. Autoimmune and allergic responses are most significant pathologically when chronic. I argued above that the primary generation of anti-inflammatory immune states following acute infections would be catastrophic and that evidence supports the suggestion that anti-inflammatory immune states are the result of chronic antigen stimulation. We are thus led to explore the idea that non-hygienic conditions, especially early in life, more often lead to anti-inflammatory immune states associated with T_{reg} cells, following some forms of chronic stimulation.

T_{reg} cells and chronic immune responses

It appears that natural T_{reg} cells are generated in the thymus, and that their specificity repertoire is biased through positive selection towards some "peripheral self antigens" expressed by specialized thymic cells. This process of positive selection appears to be quite different from the positive selection of T cells previously described. Those T cells whose TcR recognizes peptides derived from "peripheral self antigens", and whose TcR recognize such peptides presented by these specialized thymic cells, are given not only the kiss of life but also signals causing them to differentiate into natural T_{reg} cells.[122] We have noted that induced T_{reg} cells can be generated in the periphery following chronic antigen stimulation at mucosal surfaces.[123] It appears likely to me that natural and induced T_{reg} cells are similar in most respects, the major difference being their specificity repertoire. Natural T_{reg} cells are primarily directed to peripheral self antigens, and adaptive T_{reg} cells to antigens that cause substantial and chronic stimulation at mucosal surfaces.

These considerations lead to the idea that there are certain advantages to an animal in the generation of states of anti-inflammatory immunity, following some forms of chronic antigenic stimulation. The selective evolutionary forces giving rise to the regulation of immunity we now see would have occurred under less hygienic conditions than those now existing in industrialized societies. This general proposition raises three questions, which we shall now attempt to address. A consideration of these questions will allow a more defined, though frankly speculative, view as to why there is an increased incidence of Th1 autoimmunity and Th2-dependent asthma in industrialized societies.

Chronic immune responses following antigen exposure via the gut

It has long been known that chronic immunization by the oral route can give rise to the production of IgA antibody and unresponsiveness for the induction of other classes of immunity.[118] The older demonstrations of IgA immune deviation involved substantial and prolonged antigen delivery at mucosal surfaces over a considerable period. More recent observations show such immune deviation can be less extreme against gut flora, so Th1 responses can be generated against gut flora upon systemic infection. It is interesting that acute intestinal immunization with different numbers of parasites can result in acute predominant Th1 and predominant Th2 responses when a few or a relatively large number of parasites are employed for infection.[159] This dependence of the Th1/Th2 phenotype of an acute immune response parallels observations made following immunization by non-mucosal routes. These findings support an intuitively plausible scenario: acute responses to a new intestinal infection follow the rules already discussed in the context of infections and immunization by non-mucosal routes, resulting in what is usually an effective immune response to this acute infection. However, chronic Th1 responses, and to a lesser degree Th2 responses, would serve little use in attacking benign gut flora, and may lead to bystander damage that can occur following substantial, sustained, and chronic Th1 and Th2 responses. In addition, we have seen that antigens cross-reacting with self antigens have the potential of inducing CD4 T cells specific for these self antigens via the mechanism of epitope spreading, and leading to damaging Th1 and Th2 responses against these self antigens. The long-term establishment of a state of anti-inflammatory immunity to gut antigens would lead to states of anti-inflammatory immunity to the self antigens with which they crossreact, thereby minimizing damage.

In summary, it has long been recognized that there are advantages to chronic intestinal immunization leading to benign IgA immunity, mainly protective by blocking the penetration of the intestinal wall by various microorganisms, parasites, and gut flora. For example, a genetic disorder leading to IgA deficiency can lead to greater susceptibility to the parasite that causes "beaver fever", more formally known as giardiasis, due to a protozoan infection, when people drink water from a source exposed to infected beavers.[160]

Chronic immune responses following antigen exposure via the lung

The enormous load of foreign antigen in the intestine is evident. The antigenic load at other mucosal surfaces is much less, and so the evolutionary pressures might well result in different forms of regulation of immune responses. Consider antigen exposure via the lung. We have already described major findings on the nature of immunity to allergens in allergic and non-allergic individuals. They have led us to recognize that there are at least two major modes of immune response upon chronic lung exposure, namely those associated with predominant IgG_1/IgE and predominant IgG_4/IgA antibody production, reflecting the predominant generation of Th2 and T_{reg} cells. However, tuberculosis is still the most lethal infectious disease of humankind, and infection is mainly through aerosol exposure and initial trapping of the pathogen in the lung. The enormity of death caused by tuberculosis is remarkable in face of the fact that only about 5% of infected individuals become ill, and reflects the very high rate of infection. It appears that those infected but healthy individuals mount a predominant and stable Th1 response. This response contains the pathogen at such low levels that no pathological symptoms are evident, and the infection is often contained but not eliminated over decades. Such low levels of infection by the pathogens causing, for example, tuberculosis and cutaneous leishmaniasis, are apparent when the immune system is subsequently undermined, as occurs in AIDS patients, giving rise to *reactivation disease.* This is a consequence of the loss of immune control of a low-level but chronic infection, and the consequential reappearance of more substantial levels of the pathogen and of the pathological symptoms of the disease. The fact that the infection is normally kept at low levels reflects the effectiveness of the immunity generated. It makes evolutionary sense that this response does not evolve into a different mode when there is significant but minimal antigenic stimulation.

These observations are best explained by the proposal that chronic antigen impingement via lung exposure can result in at least three major types of more-or-less stable immune states. These are a Th1 response, as exemplified by the long-term state in healthy individuals exposed to *M tuberculosis* via lung infection; a predominant Th2 response, as seen in allergic individuals; and a state of anti-inflammatory immunity to the "allergen", as seen in non-allergic individuals living in a geographical area where reaction to this allergen is highly prevalent. What determines which of these three types of chronic immune response is stably established?

Chronic stimulation leading to Th1 or to Th2 responses or to states of anti-inflammatory immunity

Antigen dose and duration of stimulation critically affect the nature of the immune response that follows parental, i.e. non-mucosal, exposure. We have seen that relatively low and chronic exposure results in stable Th1 responses; chronic exposure to higher doses results in a transient Th1 response that decays as Th2 cells are generated. Increasing the dose further results in a shorter predominant Th1 phase and the more rapid generation of Th2 cells. I suppose, given the observations already referred to concerning the chronic Th1 state in healthy individuals infected with *Mycobacterium tuberculosis* via the lung, that low and chronic antigen loads can give rise to stable Th1 responses, and higher loads to Th2 responses. How does antigen load and persistence of antigen stimulation over time affect whether Th2 responses or a state of anti-inflammatory immunity is generated? I suggest for four reasons that the higher the antigen load, and the longer the stimulation, the more likely will a mucosal immune response evolve from a Th2 to an anti-inflammatory mode.

First, some younger individuals are allergic to milk antigens, but can grow out of this allergic state. If this desentisation really reflects the generation of a state of anti-inflammatory immunity, as seems likely,[161] it must mean that there can be a natural progression from a state of Th2 immunity into a state of anti-inflammatory immunity. Second, *specific immunotherapy (SIT)* for asthma/allergies consists of giving patients a series of escalating doses of antigen. This is well recognized as being something of a hazardous undertaking, as it can result in acute inflammation or *analphylaxis*, and this possibility or likelihood limits the radical application of this treatment. Nevertheless, treatment can sometimes be partially effective and, when it is, treatment results in a modulation of the Th2 response towards a state of anti-inflammatory immunity.[3] These observations also suggest that further antigen stimulation is needed to generate anti-inflammatory immune states than the stimulation needed to generate Th2 responses. Third, it seems likely that antigens of benign gut flora will be chronically present and at high amounts, and it would be advantageous if such antigens induced a state of anti-inflammatory immunity associated with benign IgA and IgG_4 antibody production. Last, the state of anti-inflammatory immunity is the least effective of the three states. It makes sense in that it is the option of last resort when chronically responding to *benign foreign or self antigens*, thus minimizing direct and bystander damage. The obvious debilitating consequence is that chronic immune responses to pathogens, against which Th1 and Th2 immune responses are insufficiently effective, may evolve into an even less effective state of anti-inflammatory immunity.

The above considerations lead me to emphasize the observations implying a major role of T_{reg} cells in helping human IgA and IgG_4 antibody responses, and suppressing other classes of immunity. The most striking observations come from the incisive studies on immune states in allergic and non-allergic people.[3,124] This line of reasoning leads us to the next question: what could be the biological role of nat T_{reg} cells?

Speculations on the biological significance of natural T_{reg} cells

I start with the likelihood that natural T_{reg} cells have a repertoire biased towards some peripheral self antigens presented by special thymic, APC-like cells.[122] I also think it likely that such natural T_{reg} cells, having strong similarities to induced T_{reg} cells in the lymphokines they produce and cell surface antigens they bear, also have a similar function in immune class regulation. It seems it would be disadvantageous if natural T_{reg} cells act to inhibit acute Th1 and Th2 responses, as they would then inhibit such protective responses against invaders. I suggest that these natural T_{reg} cells bias chronic responses to peripheral self antigens towards a state of anti-inflammatory immunity, thus minimizing damaging autoimmunity. The depletion of natural T_{reg} cells would result, according to this view, in an increase in damaging organ-specific autoimmunity, as observed.[121] I regard this proposal as an interesting but frankly speculative possibility.

Hygiene in the context of the good and the bad of T_{reg} cells

We have considered the function of natural T_{reg} cells and induced T_{reg} cells mainly from the perspective of their beneficial effects in dampening detrimental responses. However, evolution is a dialectic process and it seems most likely that natural T_{reg} cells can have harmful effects. We shall assume in our deliberations, in accord with striking evidence[120] as well as on the basis of plausibility, that antigen-specific T_{reg} cells act via the recognition of linked epitopes, so that their regulatory influence is highly specific.

Consider a situation in which the prevalence of polyclonal T_{reg} cells is much greater in a young adult than is normal. In this case, sustained Th1 and Th2 responses would be more often curtailed and to a greater extent with untoward consequences for the individual. Thus, an optimal balance between the generation of different types of CD4 T cells would be strongly selected for by evolution.

Consider the effects of changing a life style from a relatively non-hygienic to a more hygienic one. There would be less exposure to foreign antigens, leading to less chronic immune stimulation and a smaller proportion of peripheral T_{reg} cells specific for these foreign antigens. Th1 and Th2 cells would be more prevalent. It seems we are perhaps beginning to obtain a glimpse of why cell-mediated autoimmune responses and allergies are becoming more prevalent in hygienic societies.

What we do not discuss

I have now sufficiently described the advances in the field, and our own studies on how immune responses are regulated, to outline the rationale underlying our envisaged strategies to prevent or treat various undesirable and immune system-related medical conditions. However, such a description at this point would convey a mistaken impression to those not knowledgeable of immunology. We have described only a very small slice of the subject. The question arises as to whether what we have omitted to describe might not be more pertinent in judging the plausibility of the proposed strategies than the observations and concepts we have covered. Furthermore, perhaps this awareness should make us hesitant to base speculative strategies on our limited knowledge.

My reaction to this cast of mind is two fold. On the one hand, I acknowledge the tentativeness of the undertaking. On the other, my experience in witnessing how immunology has developed over the last five decades leads me to wish I had been more brave and fearless in my speculations. Clear and novel ideas are always enlightening, even if their exploration leads to an awareness that they have some limitations. The style of championing such explorations should be contingent upon an appreciation of their tentativeness. Without a realistic idea of the enormity of knowledge we apparently ignore, and without a realistic idea of the extent to which we are ignorant of nature, our considerations would appear much less perilous than they actually are. We should anticipate imperfections in our envisaged strategies or, worse, in their foundations. With this frame of mind, we can have the resolve for long journeys.

In terms of my biased account of the field, I must first acknowledge that much contemporary research is directed at elucidating the signaling pathways employed by the cells of the immune system. The immune system provides some of the best models for analyzing how mammalian cells interact and thereby affect each other's fate. These interactions involve the cells' surfaces, where signals are generated by the interaction of ligands with their receptors. These signals affect the interacting cells' expression of genes and the cell's differentiation state. Perhaps the primary aim of about a third of current immunological papers is to delineate these signaling pathways, to relate them to changes in cell behavior, and to gain an insight into how these pathways evolved. Once gained, such knowledge will allow us to recognize in what respects different cells express different genes and produce different proteins and other molecules, allowing the design of drugs that differentially affect the state of these different cell types. I have not been involved in such studies, and so this account will not describe the potential of such approaches for immunological intervention.

There are other cell types besides the B cells, CD8 T cells, and CD4 Th1, Th2 and T_{reg} cells, whose function we have partially discussed, which are important mediators and regulators of immune responses. Some families of cells bearing the Thy-1 antigen do not use the same V, D, and J genetic elements that are used to generate the genes encoding the α and β chains of the T cell receptor (TcR), but use somewhat similar but different elements that, when appropriately recombined, encode the gamma (γ) and delta (δ) chains of the TcR that these *γδ T cells* employ in their recognition of antigen. Interestingly, γδ T cells do not bear substantial amounts of CD4 or CD8 molecules on their surface; moreover, their recognition of antigen is not MHC-restricted.

There are clearly different subsets of γδ T cells in mice. The nature and prevalence of different γδ T cell subsets also varies remarkably between species. In the mouse, some γδ T cells show limited diversity, and so exist at a high frequency; this subset of γδ T cells is located at fixed sights, such as the skin. Their high frequency correlates with their lack of mobility, and hence their lack of participation in the kind of roaming surveillance that classical and scarce αβ T cells perform by circulating around the body. In addition, some of these γδ T cells appear to recognize self-proteins induced by diverse forms of stress, so-called *stress proteins*, and their recognition of antigen can lead to processes involved in wound healing. Given their highly limited diversity and their specificity, such γδ T cells most likely do not go through a tolerance process, weeding out cells with anti-self activity. The absence of a purging process to establish self-tolerance, their limited TcR diversity, their lack of mobility, and their endowment with specificity for particular self antigens distinguishes these γδ from classical αβ T cells. In contrast, other γδ T cells have properties more similar to classical αβ T cells, except that their recognition of antigen is not MHC-restricted. They constitute in mice about 5% of the peripheral pool of Thy1-bearing cells.[162] It seems likely that these distinct γδ T cell subsets, similar in some respects to α/β T cells and yet so different in others, provide clues as to how the T cells of the immune system have evolved. One could most naturally regard the non-diverse subset of γδ T cells as constituting a part of the innate defense system, and the diverse γδ T cell subset as part of the immune system.

There are in addition other types of cells that exist as different but related subsets. Some subsets can be most naturally regarded as belonging to innate defense and others to the immune system. A major family of cells is referred to as *natural killer* or *NK cells*.[163] The first subset of NK cells to be recognized was originally detected by their ability to bind to and lyse some tumor cells. Remarkably, some target cells, derived from a parental tumor line through the loss of expression of an MHC class I molecule, were found to be more susceptible to lysis by NK cells than the parental line! It became apparent that the lysis of target cells by NK cells is governed by two classes of germ-line encoded receptors: *activating receptors*, whose recognition of their ligand on the surface of the target cell promotes the lytic process, and of *inhibitory receptors*, whose recognition of their ligand inhibits the lytic process. Some inhibitory receptors recognize certain class I MHC molecules, and so a loss by target cells of these MHC molecules results in increased susceptibility to lysis. This is the basis of the *missing self hypothesis* concerning the lytic activity of NK cells.[164] Certain tumor cells, subject to CTL attack through the recognition by CTL of foreign peptides bound to the target cells' class I MHC molecules, can duck such attacks by down-regulating their expression of class I MHC molecules, or giving rise to progeny that do not express one or more class I MHC molecules. Some viruses can also down-regulate the expression of class I MHC molecules in cells they have infected, as a strategy of immune evasion. One role of NK cells appears to be a counterattack against such evasion mechanisms by target cells. The level of expression of activating and inhibitory receptors must be set to be balanced so that NK cells do not attack normal self cells, only those aberrant in some way caused, for example, by loss of expression of class I MHC molecules.

These activating and inhibitory receptors of NK cells are germ-line encoded. This subset of NK cells can be best regarded as belonging to part of innate defense. There is another subset of NK cells that has a somewhat restricted $\alpha\beta$ T cell receptor repertoire, and yet another with a more diverse set of $\alpha\beta$ TcRs. The roles of all these different cells types in regulating immune responses are not clear. However, there are some observations that prompt the suggestion that their role is not of overwhelming importance.

Mice can be generated in which the function of certain genes is abrogated or knocked-out. An examination of the physiology of such *knock-out (KO) mice* can be enlightening. Attempts to generate some of these mice seem always to result in failure. This finding is usually interpreted as implying that the gene involved is essential for survival of the developing embryo. What surprised most immunologists was that many KO mice can survive. For example, mice whose IL-2 gene has been knocked out survive, though these mice are rather sick. Mice can be generated in which certain NK T cells are absent, and they are not grossly abnormal, nor are their immune responses highly aberrant. Studies employing various knock-out mice have led us to appreciate how the functions of different molecules and cells are redundant, so that removing one cell type, or knocking-out the gene for one molecule, has a less dramatic effect than most would have originally anticipated. It therefore seems, on the basis of studies employing KO mice, that some NK T cells do not have a pivotal and essential role in regulating immune responses. In contrast, mice can be generated that are deficient in CD4 T cells. Such mice cannot generate most immune responses, implying a central role for CD4 T cells in such responses.

Synopsis of Chapter 6

We focus here on just two basic questions: what determines whether antigen activates or inactivates CD4 T cells and, if activation occurs, what determines the Th1/Th2 phenotype of the CD4 T cells generated? I first discuss considerations at the level of the system. I argue that models at the cellular and molecular level should, to be plausible, be congruent with considerations at this level.

It seems intuitively plausible that the process of activation of CD4 T cells by a foreign antigen should not generally **interfere** *with the inactivation of a CD4 T cell by a peripheral self antigen. We refer to this envisaged and cardinal feature of these processes as* **the principle of non-interference**.

The Th1/Th2 phenotype of an immune response against a pathogen is often critical to whether the pathogen is contained. It seems important that the process determining the Th1/Th2 phenotype of an immune response against one pathogen is independent of the process determining the Th1/Th2 phenotype of the immune response against another, non-crossreacting pathogen responsible for a concurrent infection. We refer to this envisaged and cardinal feature as **the principle of independence**. *We argue that the requirements for achieving* **non-interference** *and* **independence** *impact upon our considerations as to the nature of the relevant regulatory processes at the cellular and molecular level.*

We know that the activation of an anti-hapten B cell is usually only facilitated by an anti-Q T helper cell if the hapten is physically linked to Q. This requirement ensures the delivery of help by T helper cells occurs in a highly specific manner. We refer to this requirement by saying that the B cell/Th cell interaction requires the operational recognition of linked epitopes.

In considering the possibility that the presence of antigen-specific T helper cells determines whether antigen activates or inactivates naïve CD4 T cells, I felt that such a possibility would only make physiological sense if the CD4 T cell/CD4 T cell interaction, required for the activation of CD4 T cells, occurs in a manner involving the operational recognition of linked epitopes. In this case, non-interference will be achieved. I could only envisage how linked recognition could be realized if the interaction between the CD4 T cells was mediated by an antigen-specific B cell. If correct, this possibility would mean that the primary activation of CD4 T cells requires the interaction of three scarce cells.

How could this be achieved? I refer to this major problem in our ability to understand the miracles of nature as the scarcity problem.

The Two Step, Two Signal Model for CD4 T cell activation arose from several considerations. The model is consistent with a wide array of observations. It partially addresses how the scarcity problem may be overcome. The proposed mechanisms for the activation and inactivation of CD4 T helper cells achieves non-interference and is consistent with The Historical Postulate. I envisaged that antigen is presented to a naïve CD4 T cell by macrophages or dendritic cells in the first step, resulting in the proliferation of this CD4 T cell. The progeny of this CD4 T cell is envisaged in time to die unless these CD4 T cells complete the second step. The proliferation of naïve CD4 T cells in the first step means the CD4 T cells potentially participating in the second step are more frequent, hence the existence of this step, involving cellular multiplication, partially overcomes the scarcity problem. The step one primed CD4 T cells interact with antigen presented by a B cell activated by T helper cells in step 2, see Figure 35. The requirement for CD4 T cell cooperation in the activation of CD4 T cells, in the context that a direct interaction of a single CD4 T cell will result in its inactivation, provides an explanation for how peripheral CD4 T cell tolerance is achieved, in the same manner as that provided by the original Two Signal Model. Our recent in vivo *studies provide strong support for the premise that the activation of CD4 T cells, resulting in the generation of IL-2-, IFN-γ- and IL-4-producing CD4 T cells, is facilitated by CD4 T cell collaboration, mediated by a B cell as the APC.*

The Threshold Hypothesis proposes that tenuous CD4 T cell interactions, mediated by the recognition of linked epitopes, leads to the generation of Th1 cells, whereas more robust CD4 T cell interactions lead to the generation of Th2 cells. Our studies over the last forty years have tested incisive predictions unique to this hypothesis. These studies show that the combination of the amount of antigen and number of CD4 T cells determines, in the manner expected on The Threshold Hypothesis, whether primary Th1 or Th2 cells are generated. I believe the plausibility of these conclusions is considerably reinforced by the fact that parallel observations have been made in three types of systems. Some in vitro *studies employed polyclonal CD4 T cells and others employ CD4 T cells obtained from TcR transgenic mice. The second system employed lethally irradiated mice reconstituted with different types of unprimed spleen cells. We could examine in this system what cells types, as well as how many, are required to generate Thl and Th2 cells. This second system also allowed us to examine the mechanism by which the CD4 T cells interact. The mechanism demonstrated allows one to understanad how independence is achieved. The last system involved studies in intact mice. Concordant studies in diverse systems constitute convincing support for the Threshold Hypothesis. Moreover, the Threshold Hypothesis is a quantitative hypothesis, and its application brings quantitative considerations to the fore. I was stimulated by the success of our experiments in testing the hypothesis to explore whether we could develop strategies of intervention in different areas of medical interest.*

Chapter 7

THE CONCEPTUAL AND EXPERIMENTAL BASIS FOR OUR STRATEGIES OF IMMUNOLOGICAL INTERVENTION

Preface

The primary basis for our attempts to realize such strategies has been observations reported in the literature and our studies aimed at elucidating how immune responses are regulated. I wish to emphasize that this basis is simple. Our attempts have been carried out over the last twenty-five years. The conceptual framework we employ involves different and contradictory propositions to those currently highly prevalent and dominant in the field. Before I embark upon discussing our attempts, I feel it worthwhile to make it as clear as possible why I do not share these contemporary and prevalent views, and why I feel they present an impediment to progress.

It seems to me that there is currently a remarkable consensus that the activation/inactivation of CD4 T cells is well described by the PAMP/Danger Models. In addition, many immunologists agree that the nature of the PAMPs on infectious agents, or associated with an antigen, determine the Th1/Th2 phenotype of the ensuing response. Thus, most envisage that interactions of the APC with PAMPs, or with molecules associated with danger, can result in the activation of the APC such that CD4 T cells can now be activated. Moreover, many suppose that different modes of activation can lead to the generation of Th1 or Th2 cells. Alternatively, others suggest the kind of APC presenting the antigen is critical, and yet others the cytokine milieu, as already discussed. I collectively refer to these views as the ***Two Cell Framework***, as they envisage that what is critical in determining the Th1/Th2 phenotype of the effector CD4 T cells generated is the nature of the interaction of the naïve CD4 T cell with an APC, including the environment in which this two cell interaction takes place, and not on the presence/activity of other specific lymphocytes. This framework contrasts with our Two Step, Two Signal Model and the Threshold Hypothesis, according to which CD4 T cell interactions facilitate the activation of CD4 T cells and, if activation occurs, the intensity of such CD4 T cell interactions determines the Th1/Th2 phenotype of the CD4 T cells generated.

Three conceptual concerns prompt me to question The Two Cell Framework. First, in terms of the activation/inactivation of CD4 T cells, it is envisaged within this framework that whether a CD4 T cell is activated or inactivated depends upon whether or not a PAMP-dependent or danger signal occurs. This proposal implies that whether CD4 T cells are activated or inactivated by an antigen depends *only* upon the circumstances at the time of antigen impingement. This proposal violates the Historical Postulate. Second, as PAMPS and Danger signals are not antigen-specific, it is difficult to envisage how the requirements to activate a CD4 T cell by a foreign antigen will not sometimes interfere with the inactivation of CD4 T cells by peripheral self antigens. A major aim in developing the Two Step, Two Signal Model was to find and propose a mechanism for the activation and inactivation of CD4 T cells that minimizes interference, is consistent with *the Historical Postulate,* and accounts for the pertinent observations.[130] I regard our recent observations as evidence against the PAMP/Danger model for CD4 T cell activation.[136]

Third, we have described evidence for the Threshold Hypothesis, and explained how it accounts for a list of classical observations on the variables of immunization that affect the Th1/Th2 phenotype of the ensuing response. These variables include the dose of the antigen administered, as well as the time elapsing after antigen impingement, at which the Th1/Th2 phenotype of the response is assessed. The Threshold Hypothesis also accounts for the Pearson/Raffel generalization, that the degree of foreignness of an antigen is critical in determining the Th1/Th2 phenotype of the response. The Threshold Hypothesis was proposed in the early 1970s, in part because it accounted for all these quantitative observations. The Two Cell Framework not only does not account for these observations, but this framework is difficult to reconcile with them. For example, the immune response often goes through an exclusive cell-mediated Th1 phase before antibody is produced and a Th2 component of the immune response develops, as Salvin's observations and many others show, see Figure 8 in Chapter 1. However, the nature of the antigen-associated PAMPs do not change with the course of an immune response, and so PAMPs cannot determine the Th1/Th2 phenotype of the immune response as this phenotype evolves.

TABLE I

Overview of attributes of the immune system and their hypothetical basis

ATTRIBUTE of the immune system	Pertinent areas of medicine	Hypothesis as to how attribute is realized	Related considerations	Related observations
PERIPHERAL SELF-NONSELF DISCRIMINATION	autoimmunity transplantation cancer immunology	**Cell Cooperation Hypothesis for lymphocyte activation:** one lymphocyte/ multiple lymphocytes for lymphocyte inactivation/activation	Hypothesis consistent with **The Historical Postulate** as the basis for distinguishing self from non-self	Antigen inactivates B cells and CD8 T cells in the absence of T helper cells
			Principle of Non-Interference Responses to foreign antigens must not prevent the inactivation of lymphocytes specific for peripheral self antigens	
			CD4 T cell cooperation should involve the operational recognition of linked epitopes mediated by the recognition of linked epitopes. Hypothesis gives rise to the Two Signal Model of lymphocyte activation	CD4 T cell activation is facilitated by CD4 T cell cooperation
		The PAMP/Danger Hypothesis for lymphocyte activation **no PAMP/no danger: lymphocyte inactivation**	The immune system does not discriminate self from nonself, rather: infectious/danger from non-infectious/ non-danger. Hypothesis violates Historical Postulate and perhaps the Principle of Non-Interference.	PAMPs often upregulate expression of costimu- latory molecules

ATTRIBUTE of the immune system	Pertinent areas of medicine	Hypothesis as to how attribute is realized	Related considerations	Related observations
IMMUNE CLASS REGULATION	allergies cancer infectious diseases	**Threshold Hypothesis:** strength of antigen-mediated CD4 cell interactions determines class of immunity: tenuous CD4 T cell interactions: Th1, CTL; robust CD4 T cell interactions: Th2, Ab	**Principle of Independence** Critical cell-to-cell interactions should involve operational recognition of linked epitopes Meaningful connection between effector function and induction of immunity, so immunity is effective: minimization of consequences of autoimmunity Threshold Hypothesis accounts for (a) to (d)	(a) Antigens with few foreign sites can only induce Th1 cells; Pearson-Raffel generalisation (b) Partial depletion of CD4 T cells modulates Th2 to Th1 responses (c) Antigen does of very foreign antigens determines Th1/Th2 phenotype in long term. Low doses Th1 and higher doses Th2 responses. (d) Medium doses of very foreign antigens generate transient Th1 responses before mixed Th1/Th2 or Th2
		The PAMP Hypothesis: The nature of the PAMPS determines the Th1/Th2 phenotype of the immune response generated	Pathogen has control over class of immunity generated PAMP Hypothesis is inconsistent with (a)-(d) above	PAMPs can upregulate costimulatory molecules and some cytokine production: does/may influence Th1/Th2 phenotype of response

It is difficult to see how the Two Cell Framework could account for the antigen dose-dependence of the Th1/Th2 phenotype of immune responses and for low-zone and high-zone cell-mediated immune deviation. In addition, we have shown that the Th1/Th2 phenotype of an immune response, generated *in vitro* or *in vivo*, depends on the number of CD4 T cells present when the amount and nature of the antigen is not changed. These findings on the dependence of the Th1/Th2 phenotype of the response on CD4 T cell number cannot be accommodated within the Two Cell Framework. According to it, decreasing the number of CD4 T cells should just decrease the size of the immune response generated and not affect its Th1/Th2 phenotype. Last, the PAMP Model endows the pathogen with control over the nature of the response generated, in the sense that its genome controls the PAMPs expressed. This possibility makes at least me uncomfortable.

A comparison of how the two conceptual frameworks are related to various considerations at the level of the system, and to various observations, is given in Table I.

We draw two simple conclusions from our basic studies concerning how the Th1/Th2 phenotype of an immune response is determined. These conclusions have provided the driving force for our studies on immunity to infections and against cancers. It is helpful to state upfront what these two tentative conclusions are. Consider a situation where the administration of antigen results in a predominant Th2 response. Reducing either the dose of antigen or the number of responding CD4 T cells will change the Th1/Th2 phenotype of the ensuing response, increasing the Th1 component at the expense of the Th2 component. Naturally, the degree of the decrease in the number of CD4 T cells or in the amount of antigen must be appropriate. In addition, a combination of reducing the antigen dose and number of CD4 T cells can be effective. We will show such manoeuvres are effective in beneficially modulating primary and ongoing immune responses. We have therefore found these two conclusions to be fruitful in achieving significant aims, as I outline below.

The development of our low dose vaccination strategy

When I was at the Salk Institute in the early 1970s, a number of us younger members, graduate students and postdoctoral fellows, who were doing studies in basic immunology, felt our research should be more directly concerned with the relief of human suffering. Our social conscience was heightened by the Vietnam War and our reaction to the values expressed by the Nixon White House. Donato Cioli, Paul Knopf, Alan Sher, and myself constituted this group. These three friends all entered the field of schistosomiasis, changing their field of research dramatically. I was tempted to change direction also but felt, on reflection, that for me the time was not yet ripe. However, I subsequently tried to keep abreast of major developments in the broad field of immunoparasitology.

In the early 1980s, I and an especially gifted technician, Mohamed Dhalla, were developing the *in vivo* and *in vitro* systems with which we could test predictions of the Threshold Hypothesis. I received at this time a letter from Graham Mitchell, following the publication of our first *in vivo* studies directed at testing this hypothesis. Graham had undertaken with Jacques Miller some classical immunological studies. These studies enlarged upon Henry Claman's initial finding that the generation of antibody responses requires cooperation between bone marrow and thymus cells, and Graham showed the antibody-producing cells were descended from the bone marrow cell. Graham had also completely changed the subject of his research to immunity against parasites. He wrote to me because the studies Emanuela

Handman and he had undertaken in the mouse model of cutaneous leishmaniasis seemed so parallel to ours, whose aim was to understand how the class of immunity is determined. I will give an account, in contemporary terms, of what was then known about the murine model of cutaneous leishmaniasis, as this is necessary to appreciate our mutual excitement, and to understand the scientific significance of our complementary findings.

It was known from studies of infected human subjects and experimental observations in infected animals that infection by various intracellular pathogens, able to cause chronic/progressive disease, did not invariably lead to illness. These pathogens include those responsible for AIDS, tuberculosis, and the leishmaniases. The latter group of diseases is caused by different species of the protozoan parasite *Leishmania*. The species *Leishmania major* grows best at a slightly lower temperature than 37°C, and so grows optimally just under our skin, subcutaneously, and can cause **cutaneous leishmaniasis**. These leishmania parasites grow inside macrophages. Infection of humans, occurring through the bite of an infected sand fly, can lead to two extreme outcomes: a local and contained lesion that occurs at the site of the initial infection that with time resolves with no untoward consequences, or to metastasis of the parasite from the original site of infection and the formation of distal lesions. Studies in humans have led to the recognition that infections leading to parasite containment are associated with a predominant Th1 response, and that the immune response of patients has a substantial Th2-component. Interestingly, when James Howard developed these mouse models of cutaneous leishmaniasis, he made parallel findings. When different strains of mice are given a standard challenge of 10^6 parasites subcutaneously, some develop a transient lesion at the injection site, which resolves with time, whereas in mice from other strains there is metastatic spread of the parasite to form lesions distal to the injection site. Strains that contain the parasite were designated as *resistant*, and those suffering metastatic spread of the parasite were designated as *susceptible*. Interestingly, resistant mice produce a predominant Th1 response on standard parasite challenge, whereas susceptible mice mount with time a predominant Th2 response.

The most susceptible mouse strain is a white strain known as **BALB/c**. Thus BALB/c mice, when infected with 10^6 parasites according to the standard protocol, mount a predominant Th2 response associated with parasite spread, including spread into some internal organs such as the spleen. Emanuela Handman and Graham Mitchell had made use of a BALB/c mouse that had a mutation, precluding the development of T cells and also, as it happens, of fur. These mice are naked and so are known as nude BALB/c mice. As expected, nude BALB/c mice are immunoincompetent. Emanuela and Graham had found that these mice, reconstituted with a normal complement of CD4 T cells of about 5×10^7, and infected with a standard challenge of *Leishmania major*, were susceptible like normal BALB/c mice and mounted a similar type of immune response. The real surprise was their finding that nude BALB/c mice, reconstituted with much fewer CD4 T cells, only a few million, and given the standard challenge of a million parasites, were resistant. Moreover, they made a similar anti-parasite immune response as mice of resistant strains.[165] Graham wrote to me as he recognized the similarity of their findings to ours, shown in Figures 39 and 40 of Chapter 6, in our studies attempting to examine how the Th1/Th2 phenotype of an anti-SRBC immune response is determined. The observations obtained in both systems were expected on the Threshold Hypothesis. I eagerly studied the literature on the mouse model of cutaneous leishmaniasis. I wished at this time that I had funds to embark upon studies in this mouse model of an important and large group of human infectious diseases.

I joined the faculty of the University of Saskatchewan in 1987. The province of Saskatchewan had a program whereby new faculty could apply for a small grant to start up a research program in a new area. I knew what I would do. I would apply for studies employing the mouse model of cutaneous leishmaniasis and, if my application was successful, would embark on such studies. I had previously felt I could not gamble on using my limited research funds, acquired to do studies on the mechanisms of immune regulation, to initiate research in a field entirely new to me. I was also thrust into a full load of undergraduate teaching when I arrived at the university, and pretty regularly got up around three in the morning to prepare my lectures. I was awarded the grant, but postponed starting up this research program for a couple of years, as I could not handle this new endeavor in addition to the effort I was spending on my lectures and on maintaining my other research.

The time came when we tried to establish the murine model of cutaneous leishmaniasis. The first experiments were done together with a colleague, Guojian Wei. I had decided we should infect different susceptible BALB/c mice with different numbers of parasites, starting from the standard number of 10^6 and going down in increments to a hundred, and monitor lesion development, see Figure 43. This experiment was motivated by the studies of Salvin and Chris Parish on how the dose of non-living antigens affects the Th1/Th2 phenotype of the ensuing immune response, with the hope there might be a similar dependence for living organisms that multiply slowly. We made several provocative findings in this experiment.[166] First, it had been shown that, when mice are injected subcutaneously with parasites into a footpad, the subsequent swelling of the foot is often a reasonable measure of the parasite burden. We found, as others had shown, that the feet of mice injected with 10^6 parasites increased in size progressively, in a pattern that had been identified as indicating progressive disease. However, the course of disease, as indicated by changes in foot size, was much more variable in mice injected with 10^3 or with 10^4 parasites. In many cases, the size of the injected foot increased and then remained swollen to a constant degree.

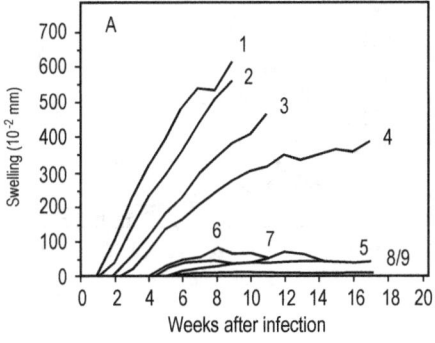
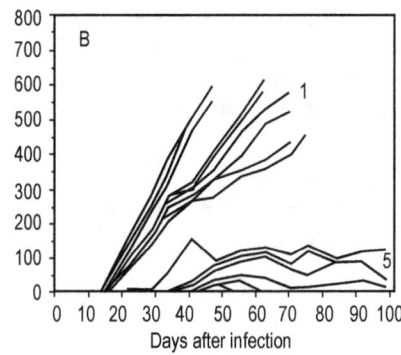

Fig. 43. Dependence of lesion size on number of *Leishmania major* injected. **A** Average lesion size following injections with different numbers of parasites. Challenges 1-9 of panel A consisted of 10^6, $3.3x10^5$, 10^5; $3.3x10^4$, 10^4; $3.3x10^3$; 10^3; $3.3x10^2$, and 10^2 parasites respectively. **B** Lesion size of individual mice injected with 10^6 parasites (1) or $3.3x10^3$ parasites (5)

Most interestingly, the size of the foot in some cases first increased in size and then returned to a normal state. This appeared to be like lesion formation in resistant individuals in that the initial lesion resolves. The feet of mice injected with 10^2 parasites rarely showed any change in size. We were eventually able to show that the nature of the long-term immune response against the parasite depends upon the initial size of the inoculum injected: high numbers of parasites, around 10^6, led to predominant Th2 responses, about $3x10^3$ led to mixed Th1/Th2 responses and chronic but stable lesions, and 10^2 parasites led to stable and predominant Th1 responses. We drew the inference that the susceptibility of BALB/c mice was not an intrinsic trait, but is dependent on the size of the parasite challenge.

The experiment we thought most interesting, however, was whether we could cause low-zone cell-mediated immune deviation by infection with low numbers of parasites and thereby make such mice resistant to a challenge with a high number of parasites, which led to progressive disease in normal mice. We challenged normal BALB/c mice, and BALB/c mice that had been exposed to either 10^2 or 10^3 parasites two months previously, with 10^6 parasites, and examined the lesion size generated, see Figure 44. Pre-exposure to low numbers of parasites made mice resistant to a high dose challenge.

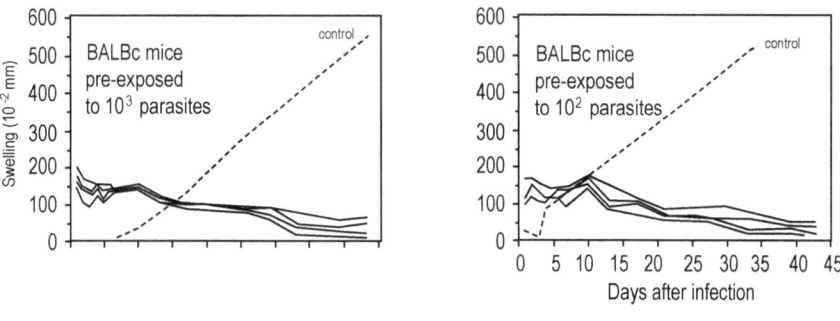

Fig. 44. Pre-exposure by infection with a low number of parasites renders "susceptible" BALB/c mice resistant to a high dose challenge. Adapted from reference 166

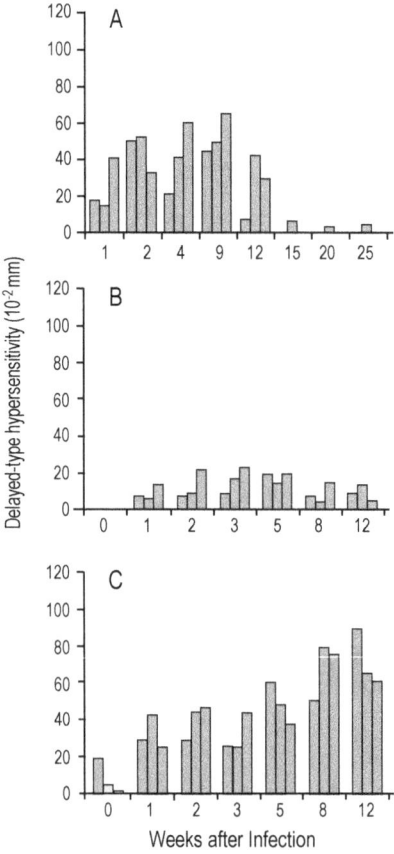

Fig. 45. Anti-parasite DTH responses in naïve mice infected with a low number of parasites (3.3×10^2), panel A: a high number of parasites (10^6), panel B and in mice pre-exposed to 3.3×10^2 parasites and subsequently challenged with 10^6 parasites, panel C. DTH responses of individual mice, three per group, are shown. Adapted from reference 166

We next examined whether this resistance, following low dose vaccination, was truly due to the deviation of the immune response from a humoral to a cell-mediated mode, and found this to be the case. Thus the DTH response to parasite antigens is low and transient in naïve mice infected with 10^6 parasites, more substantial and prolonged in mice infected with 3.3×10^2 parasites, and strongest and most sustained in mice pre-exposed to 3.3×10^2 parasites and then challenged with 10^6 parasites, see Figure 45. We refer to the principles underlying this means of achieving resistance as the *Low Dose Vaccination Strategy*. We were lucky that the manuscript reporting these observations was accepted by the leading journal *Science*. We ended our paper with the suggestion that this Low Dose Vaccination Strategy might be pertinent to efficacious vaccination against tuberculosis and against AIDS.[166]

We were all excited by these findings. Speaking for myself, I could not believe how dramatic and simple our strategy was. I sometimes became exhausted as I repeatedly reflected on the implications of this study for preventing a whole series of infectious diseases, including AIDS and tuberculosis. These thoughts encouraged me to think of further barriers to successful vaccination, and how they might be overcome.

The problem posed by genetic diversity in achieving universally efficacious vaccination

A recognized barrier to developing a vaccination strategy, universally effective among people, is the genetic diversity of the human population. We know that vaccination of people or animals with a standard protocol usually results in diverse immune responses, both in their qualitative nature, such as their Th1/Th2 phenotype, and in their intensity, due to genetic differences between individuals. Given the extent of this genetic diversity, is it hopeless to imagine that a standard and universally efficacious protocol of vaccination is achievable?

It seemed to me that, if such a vaccination strategy was ever to be realized, it must reflect rules that govern the regulation of immune responses in all, or the large majority, of people or animals. The murine model of cutaneous leishmaniasis appeared to be a good system to test any ideas one might have on achieving universally effective vaccination, as the nature of the immune response to a standard challenge of 10^6 parasites was known to vary dramatically in genetically different mice. What rule might provide a basis for developing a universally efficacious vaccination strategy?

Host	Parasite Strain	Route	Dose	S/R
"Susceptible"	L. major	s.c.	5×10^2	R
BALB/c		s.c.	1×10^6	S
"Susceptible"	NIH 173	s.c.	33	R
BALB/c		s.c.	1×10^4	S
"Resistant"	NIH 173	s.c.	1×10^4	R
CBA		s.c.	5×10^8	S
"Intermediate"	NIH 173	s.c.	33	R
A/J		s.c.	5×10^7	S
"Resistant"	NIH 173	i.v.	33	R
CBA		i.v.	1×10^8	S

Table II. Mouse strain, parasite strain, and route of infection all affect whether mice resist or succumb to a parasite infection, but in all cases mice resist a relatively low and succumb to a relatively high challenge. R and S reflect a state of resistance and susceptibility. The groups as listed correspond to the panels of Figure 45 denoted as A through E. Adapted from reference 167

In pondering this question, I was influenced by the idea that the antigen dose-dependence of the Th1/Th2 phenotype of the immune response is probably not an incidental finding in a few systems; if the Threshold Hypothesis is by and large correct, it had to be an almost universal rule. I thought there could be genetic differences between mouse strains that determine the number of parasite-specific CD4 T cells in naïve mice, and other host genes that affect, for example, the rate at which this intracellular parasite multiplies and thereby controls the effective antigen load following infection. Thus, a number of parasites that is considered low in one strain, giving rise to a sustained Th1 response, might well be assessed as high in another strain, as infection with this number of parasites generates a predominant Th2 response and progressive disease. We therefore infected different strains of mice with different strains of parasites, sometimes even by different routes, such as subcutaneously or intravenously, with different numbers of parasites. In all cases examined, we found that relatively low numbers of parasites did not cause disease and relatively high numbers did, see Table II. We could in this way define, for a given host, a given parasite strain, and a given route of infection, a **_transition number_** of parasites: infection with a number of parasites below the transition number resulted in resistance, and infection above the transition number resulted in progressive parasite growth.[167] Thus the resistant CBA strain required infection with 5×10^8 parasites of the standard strain of *Leishmania major*, to suffer progressive infection, whereas this number was about 5×10^3 for BALB/c mice, for the same parasite strain when infected at the same site. Thus, genetics in this case accounts for a 10^5-fold difference in the transition number characteristic of these two mouse strains.

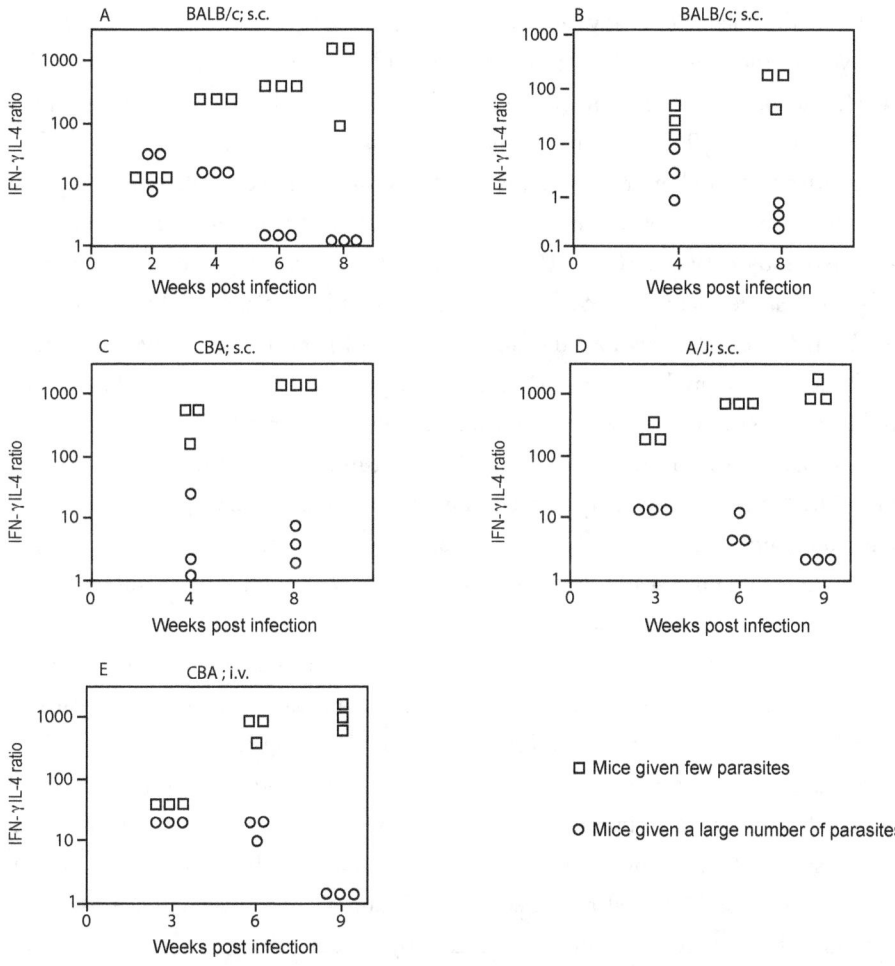

Fig. 46. The ratio of IFN-γ to IL-4 production by parasite-specific Th cells at different times after mice are injected with relatively low and high numbers of parasites. Panels A-E indicate different systems in which the host, the parasite or route of infection, are different, as listed in Table II of the previous page. Adapted from reference 166

We then examined, under these diverse circumstances, whether infection with a relatively high number of parasites, which causes progressive disease, induces a predominant Th2 response, whereas infection with a low number, which does not cause progressive disease, generates a stable Th1 response. The truth of this proposition is evident from the observations recorded in Figure 45. We infected mice of a given strain with a relatively low and others with relatively high number of parasites, so the infection was respectively contained or resulted in progressive disease. We then measured the production of IFN-γ by parasite-specific Th1 cells and of IL-4 by parasite-specific Th2 cells, present in the spleen, upon acute stimulation with parasite antigens. We assessed in this manner the relative presence of parasite specific Th1 and Th2 splenic cells at various times after infection.

We plot the ratio of IFN-γ to IL-4 produced by these parasitic-specific Th cells. It is low at eight weeks post-infection for infections with a high number of parasites, as IL-4-producing Th2 cells predominate, whereas this ratio is high following infection with a low number of parasites, as IFN-γ-producing Th cells still predominate at this time. About eight weeks post-infection, the difference in the value of this ratio is consistently between 100 to 1000 in mice infected with high and low numbers of parasites, see Figure 46. We have thus established what appears to be a widely applicable rule: relatively low and high numbers of parasites respectively generate in the long-term predominant Th1 and Th2 responses. This universal rule, if valid, holds promise of providing the means for achieving universally efficacious vaccination. I illustrate the proposed means of achieving this aim, by overcoming the barrier posed by the genetic diversity of the population to be immunized, in terms of vaccination against tuberculosis. Several proposals to the Canadian Institutes of Health Research to obtain research funds to test this strategy were unsuccessful, which is why these studies have not advanced.

Worldwide, tuberculosis is still the leading cause of death from an infectious disease, and is responsible for 2-3 million deaths annually. It is primarily caused by a bacterium, *Mycobacterium tuberculosis*, which grows inside macrophages. This mycobacterium does not produce toxins and is itself nontoxic. The pathology of tuberculosis is due to an immune response against infected cells. When the immune response is not sufficiently strong or effective to contain the initial infection, the load of bacterially-infected cells becomes substantial. The often intense immune response that subsequently develops causes bystander damage, most often in the lung, through the formation of large lesions called granulomas. Many reviews on the subject of tuberculosis describe diverse means by which *Mycobacterium tuberculosis* can evade innate resistance or adaptive immunity. The existence of prevalent and effective mechanisms of evasion would provide grounds for pessimism that this disease can ever be controlled. These judgments should be tempered in light of the fact that 90-95% of people infected with *M tuberculosis* do not become ill because the combination of their innate mechanisms of defense and their immune response contains the infection. This fact cannot but make one optimistic that an effective vaccine can be developed.

I will consider how universally efficacious vaccination against tuberculosis might be achieved within the framework that establishing a *Mycobacterium tuberculosis* specific Th1 imprint, in everyone, will protect all from tuberculosis. I develop arguments for the plausibility of this framework in the next chapter.

There is a vaccine against tuberculosis. French microbiologists Camille and Guéron cultured for years the pathogen that causes cattle tuberculosis, *Mycobacterium bovis*. They thereby developed an attenuated strain that is no longer pathogenic in cattle. This **Bacille Camille Guéron**, or **BCG**, has been used as a vaccine in attempts to protect humans and cattle against tuberculosis. It has been employed worldwide more than any other vaccine. It can clearly provide protection as assessed in some trials.[21] A big trial was undertaken in the late 1960s and early 1970s involving over a quarter of a million subjects, the famous Madras Trial held in India. Those vaccinated in this trial were, overall, as susceptible as unvaccinated individuals to tuberculosis, and so the efficacy of protection provided by vaccination was around 0%.[168] This result does not mean vaccination had no effect. It is most likely that vaccination made some individuals more resistant, others more susceptible, and did not affect the resistance/susceptibility of yet others. I shall shortly return to the remarkable result of this trial.

Consider three genetically different individuals that are to be vaccinated, all with the same dose of the antigen and by the same route. The long-term nature of the immune response expected in these individuals, when immunized with different amounts of antigen is different, as their "transition number" is different, as indicated in Figure 47. These three individuals are anticipated to often produce different types of response on immunization with a standard amount of antigen. Consider what happens if we immunize with an amount of bacterial antigen indicated by A in the figure. All three individuals will produce an inappropriate antibody response. If we chose an amount of bacterial antigen represented by B, we will generate an appropriate Th1, cell-mediated response in individuals 2 and 3, but an inappropriate antibody response in individual 1. Administration of dose C induces an appropriate immune response in individuals 1 and 3, but induces no response in individual 2.

Suppose now that we immunize not with bacterial antigens, but with a slowly growing and attenuated bacterium such as BCG, and that the indicated immune response is what would be generated two weeks after infection by the BCG. Consider the consequences of administering the dose D. Initially, an immune response will only be generated in individual 1, and it will be an appropriate Th1 cell-mediated response. The growth of the inoculated bacteria given to individuals 2 and 3 will not be restrained by any immune response two weeks post infection, and so the bacteria will grow until they reach a load corresponding to the cell-mediated zone, whereupon they will generate the desirable Th1 cell-mediated response. Thus, infection of the three individuals with a number of mycobacteria corresponding to the dose D will generate Th1 responses and Th1 imprints in all three individuals. As an aside, there are references in the older literature to a million BCG as constituting a low infection for mice. We have found that young BALB/c mice can respond well to infections with as few as forty BCG, the lowest challenge we assessed. These challenges generated robust Th1 responses and Th1 imprints.[169, 170]

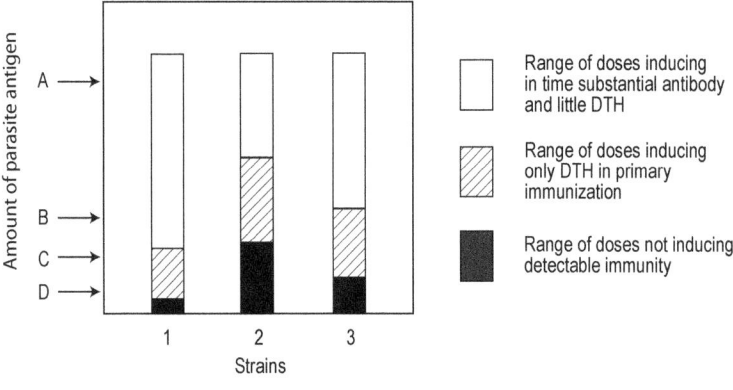

Fig. 47. Illustration of how the class of immunity is expected to depend on antigen dose in three genetically diverse individuals, and how the facts illustrated can be employed to devise a universally efficacious vaccination strategy.

These considerations can be enlarged upon and encapsulated by trying to formulate a vaccination strategy that holds the promise of providing universal protection against tuberculosis. First, any rational strategy must identify the immunological parameters that discriminate between the immunity that leads to protection, and present in healthy infected individuals, and the immunity that leads to disease. I shall argue in the next chapter that an insufficiently strong Th1 cell-mediated response, or a response with a detrimental Th2 component, leads to disease, whereas healthy infected individuals generate a sufficiently strong Th1 response to contain the infection shortly after infection. I also suggest later that the generation of a BCG-specific Th1 imprint will guarantee a strong and predominant Th1 response upon natural infection with *Mycobacterium tuberculosis* in all individuals vaccinated, and so contain the infection. How can such a Th1 imprint be achieved?

We suggest approaching this by employing the strategy of low dose vaccination. How can the problem posed by the genetic diversity of the human population be addressed? Vaccination with an ultra-low dose of replicating BCG, which does not generate an antibody response in any of the individuals, is expected in time to stimulate a Th1 response and generate a Th1 imprint in all individuals.[171]

Do these considerations bear at all on the failure of the Madras BCG vaccination trial to provide significant protection against tuberculosis? There was little rationale at the time of this trial for choosing the dose of BCG. In the absence of such knowledge, the highest dose was chosen that was known to have minimal side affects.[168] It is interesting in this context that, in several trials of BCG vaccination of cattle, no protection against cattle tuberculosis was achieved. Bryce Buddle and his colleagues in New Zealand appear to have shown that considerable protection against *Mycobacterium bovis* can be achieved by reducing the dose of BCG administered by around a *million-fold*.[172]

Developments in the murine model of cutaneous leishmaniasis

The mouse model of cutaneous leishmaniasis has become the most studied and developed model for those human infectious diseases in which predominant and strong cell-mediated responses are associated with protection, and progressive disease with immune responses that have a substantial Th2 component. I will give a partial account of the major findings made in this model for their own interest and to provide a context for a description of our further studies.

Various manoeuvres, around the time of infection of susceptible BALB/c mice with a million parasites, which normally results in progressive disease, prevent disease from developing. Moreover, an examination of the immune response in all these cases shows that these manoeuvres prevent the development of a predominant Th2 response, with a predominant Th1 response taking its place. Such observations provide strong support for the proposition that the Th1/Th2 phenotype of the response determines the outcome of infection. In addition, the nature of such manoeuvres is of great interest in shedding light on how the Th1/Th2 phenotype of the response is determined.

One manoeuvre was independently discovered by different investigators, and might have been anticipated from the earlier findings of Emanuela Handman and Graham Mitchell. [165] We have noted above that they had shown that nude BALB/c mice, which were reconstituted to have far fewer CD4 T cells than a normal BALB/c mouse, resist a challenge of a million parasites. Investigators many years later found that administering anti-CD4 antibodies to susceptible BALB/c mice, around the time of infection with a million parasites, resulted both in a considerable depletion of CD4 T cells and prevented progressive disease.[173] The effectiveness of this manoeuvre makes sense in terms of the Threshold Hypothesis.

A second manoeuvre, as already outlined, is the administration of antibody that neutralizes the activity of IL-4. We have seen that IL-4 is an autocrine lymphokine, in that it is made by Th2 cells and stimulates these Th2 cells to divide. Neutralising the activity of IL-4 thus undermines Th2 cell expansion as well as the inhibition by Th2 cells of the generation of Th1 cells. The early transition from an initial Th1 to a mixed Th1/Th2 phase, which is the normal pattern in susceptible mice, is thus undermined, and so the generation of Th1 cells is sustained.[174]

These two manoeuvres result in a predominant Th1 response if administered within days of infecting BALB/c mice with 10^6 parasites. Without these manoeuvres, a predominant Th2 response is established in about a week, and it was found that these manoeuvres are then no longer effective in modulating the immune response and in preventing progressive disease. It thus appeared that these manoeuvres are ineffective as a treatment once disease is established.

Studies directed at curing leishmaniasis in the mouse model

I had some reservations concerning the argument that these manoeuvres, of administering antibodies that either temporarily deplete CD4 T cells or temporarily neutralize IL-4, could not cure established disease. The immune response of BALB/c mice, following infection with a million parasites, becomes heavily polarized soon after infection into an exclusive Th2 mode. Such extreme polarization is rare in human disease. For example, it appears in a rare form of tuberculosis, called miliary tuberculosis, which, if untreated, has an unusual and precipitous course resulting in rapid death. Most tuberculosis is chronic and associated with the expression of cell-mediated immunity in the form of DTH. This is why a positive PPD skin test is a clinical indication for tuberculosis, as previously discussed. Most human disease is associated with a mixed Th1/Th2 rather than a heavily polarized Th2 response. I thought that, under circumstances where there is a mixed Th1/Th2 response, these manoeuvres might well be effective in curing established disease, as there would then most probably be a continuous battle between Th1 and Th2 cells. It might be easier to tip the balance towards a Th1 pole in such situations than when there is a predominant Th2 response. However, these experiments to cure mice of cutaneous leishmaniasis seemed to me not so enlightening from the point of view of understanding basic processes, so I put these thoughts on a back burner.

As our studies in the murine model of cutaneous leishmaniasis progressed, I became interested in other infectious diseases, and became motivated to explore the possibility of trials to ascertain the validity of the Ultra-Low Dose BCG Vaccination Strategy against tuberculosis. These interests led me to visit Ethiopia a few times, in the mid to late 1990s, as a guest of the Armauer Hansen Institute, situated at a high elevation on the outskirts of the Ethiopian capital, Addis Ababa. The institute was named after the Norwegian who had discovered that *Mycobacterium leprae* is the cause of leprosy, and the institute was jointly funded by the Ethiopian, Norwegian, and Swedish governments. The director, Sven Britton, made me feel most welcome. I also visited Nairobi, Kenya, to attend discussions on AIDS. I was confronted on these visits with seeing people with AIDS. I believe that there are substantial observations supporting the view that HIV can be contained by a sufficiently strong, cell-mediated, CTL anti-viral response, and that disease progression is similar to cutaneous leishmaniasis in becoming progressive if a substantial Th2 component of the response develops. I came back from a trip to Nairobi with a new resolve after seeing quite a few advanced AIDS patients. I decided we had to explore the possibility of developing treatments for these diseases.

The approach we took was straightforward. We knew that infecting BALB/c mice with about 10^2 leishmania parasites led to a predominant Th1 response and parasite containment, whereas infection with 10^6 quickly led to a predominant Th2 response and progressive disease. Infection with about 10^4 parasites led to semi-stable and chronic lesions, see Figure 42, a situation I believed was much closer to most clinical states seen in humans. The anti-parasite immune response associated with the formation of such chronic lesions had a mixed Th1/Th2 phenotype. Jude Uzonna, a postdoctoral fellow in my laboratory, showed that partial depletion of CD4 T cells, or the administration of antibody capable of neutralizing IL-4, resulted in both a rapid modulation of the response from a mixed Th1/Th2 to a Th1 phenotype and a resolution of the lesion. Moreover, after this modulation, the mice resisted a challenge of a million parasites.[175] This study holds promise that immunotherapy can be realized, perhaps in people at early stages of AIDS, as I discuss in Chapter 8.[6]

The immunological basis underlying the curing of human visceral leishmaniasis

Visceral leishmaniasis, caused by *Leishmania donovani*, is lethal unless treated by administering anti-parasitic drugs. These drugs are highly toxic for humans and there is significant morbidity during treatment. It is a remarkable fact that successfully treated individuals, even when living in an endemic area, rarely if ever become ill a second time.

Many people had shown, including ourselves, that the IgG isotype of the anti-parasitic antibody produced by mice reflected the Th1/Th2 phenotype of the associated immune response. Mice with an exclusive Th1 response produced undetectable antibody, whereas mice with a merely predominant Th1 response predominantly produce IgG_{2a} parasite-specific antibody. Mice with a predominant Th2 response predominantly produce IgG_1 antibody, whereas mice with a mixed Th1/Th2 response produce a mixture of IgG_{2a} and IgG_1 antibodies. It was not a mystery as to why these associations occur. The production of IFN-γ by CD4 T cells was known to be able to enhance IgG_{2a} antibody responses, and of IL-4 IgG_1 antibody responses. These correlations allow one to monitor indirectly the Th1/Th2 phenotype of a response against an antigen by monitoring the relative prevalence of the IgG isotypes of the IgG antibody produced against the antigen.

There are technical problems in assessing in humans the Th1/Th2 phenotype of immune responses. Juthika Menon, who as a PhD research student had done some of the critical vaccination studies in the leishmania model, later tried to repeat some marginal observations reported in the literature over a period of two years. This research was aimed at defining differences in the Th1/Th2 phenotype of the immune response against *M. tuberculosis* of healthy people, infected with the pathogen, and of tuberculosis patients. These attempts on our part were unsuccessful. In the end, we decided to look at the relative prevalence of IgG isotypes among the mycobacterium-specific antibodies as a surrogate marker for the Th1/Th2 phenotype of the anti-mycobacterium immune response. I shall describe these studies in greater detail in Chapter 8. The observations made gave further credence to inferences we had tentatively drawn from the literature that predominant Th1 responses are associated in humans with production of IgG_2 antibody and predominant Th2 responses with IgG_1 antibody. Such IgG isotype analysis is simpler technically than the direct assessment of the Th1/Th2 phenotype of immune responses.

We had come to know a medical researcher, Asrat Hailu, on our visits to Ethiopia, who worked at the Institute of Pathobiology in Addis Ababa. Asrat had been studying people exposed to *Leishmania donovani* in a village where visceral leishmaniasis was endemic. Asrat had the foresight to take serum samples to ascertain who was **seropositive,** i.e. to find people who had significant antibody levels to the parasite, an indication of infection. Asrat had collected serum samples from patients, both before and after drug treatment, and also from healthy infected individuals. It was already known from the literature that healthy contacts had a predominant cell-mediated response, whereas patients did not express, or expressed barely detectable, DTH, but produced substantial antibody. We examined the IgG isotypes of the antibody specific for leishmanial antigens in healthy infected individuals, in patients before and after drug treatment, and in normal individuals. There was little antibody to leishmanial antigens in normal people, so these observations are omitted from Figure 48.

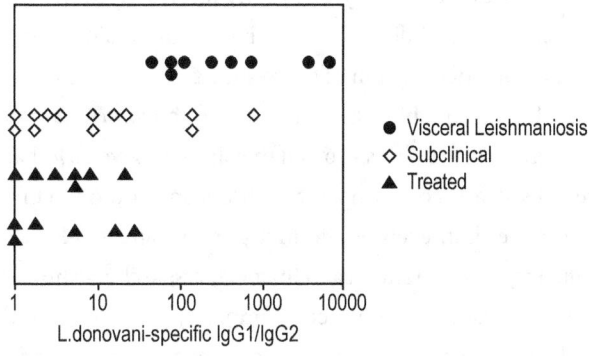

Fig. 48. The ratio of IgG_1/IgG_2 isotypes in the IgG antibodies specific for *Leishmania donovani* of individual healthy infected and of individual patients before and after drug treatment. Adapted from reference 176

Two inferences can be drawn from our findings and excited me considerably. The IgG isotype analysis showed that the value of the IgG_1/IgG_2 ratio of healthy infected and of patients barely overlapped, and so could be useful in distinguishing differences in the state of immunity between these individuals. The individual, clinically diagnosed as a healthy infected and who had the highest IgG_1/IgG_2 ratio of all healthy infected studied, whose IgG_1/IgG_2 ratio is shown in Figure 48 as that of a healthy infected individual, was the only one to subsequently come down with symptoms of disease.[176] This gave us even more confidence in the use of the IgG_1/IgG_2 ratio as an indicator of whether the immune response was effective in containing the pathogen. Second, the ratios of IgG isotypes of drug-treated patients and healthy infected individuals were indistinguishable. This finding provided a natural explanation for why patients, successfully treated with drugs, were resistant to re-infection, and further demonstrated that human immune responses could be modulated backwards, from what we inferred was a mixed Th1/Th2 to a Th1 phenotype. I regard this as a most important conclusion.[176] We shall discuss later how it might be pertinent to the design of strategies to reverse the early stages of AIDS and in cancer therapy.

Advances in the field of tumor immunology

I have been attempting to keep abreast of the important developments in cancer immunology over the last almost fifty years. I have already noted that studies in the 1960s gave rise to the view that cell-mediated immunity was protective against cancers, and it was widely believed at this time, based on observations, that the production of antibody was associated with progressive tumor growth.[67, 68] I have described how these generalizations contributed to my formulation of a Theory of

Immune Class Regulation in the early 1970s. My interest in tumor immunology has led us to some recent studies. I describe more recent and major advances in the field to provide a context for a description of these studies.

In the 1950s, the remarkable phenomenon of ***concomitant immunity*** against tumors was discovered. Researchers showed that animals given a lethal dose of tumor cells could often resist a second implant of the same tumor, typically implanted about a week after the first, even as the first grew progressively. This observation appears at first glance to be paradoxical. However, researchers showed that animals, at the time when they could resist a second implant, had T cells in their lymphoid organs that could protect a normal syngeneic animal against a normally lethal tumor challenge. Perhaps the first implant is not rejected because it has grown considerably since being implanted, and so some of the tumor cells are less accessible to host lymphocytes than the tumor cells in a much smaller tumor mass. Whatever the case might be, the transfer studies establish beyond doubt that the immune system of mice, given a lethal implant of a tumor, often produce T cells that can protect against this tumor. What then prevents these protective cells from containing the tumor?

Robert North spent years addressing this question in mouse models of cancer, employing transplantable tumors. He wrote long and insightful reviews in the 1980s, describing not only his studies, but justifying an overall view of what normally occurs when mice are given a lethal challenge of tumor cells.[177] His studies influenced me enormously. He summarized his findings in the manner shown in Figure 49. Note that tumor immunologists usually inject the transplantable tumor into a site close to the surface of the animal's body, so the size of the tumor can be readily measured. North showed that, when mice are given a lethal challenge of most tumors, protective T cells are generated. He associated these protective cells with cytotoxic T lymphocytes, or CTL, that can lyse the tumor cells. These CTL obviously do not completely contain the tumor, as the dose of tumor cells has been chosen to cause progressive tumor growth, but it is a fact that net tumor growth is often partially arrested at the time that North showed concomitant immunity is maximally expressed; this can be inferred from the slight undulation of the line of Figure 49, representing tumor growth, around the time concomitant immunity peaks. North and his colleagues showed that this peak expression of concomitant immunity is followed by the appearance of CD4 T cells that can suppress the generation/activity of the CTL. North summarized his resolution of the paradox of progressive tumor growth and the existence of concomitant immunity as "too little (concomitant immunity) too late."

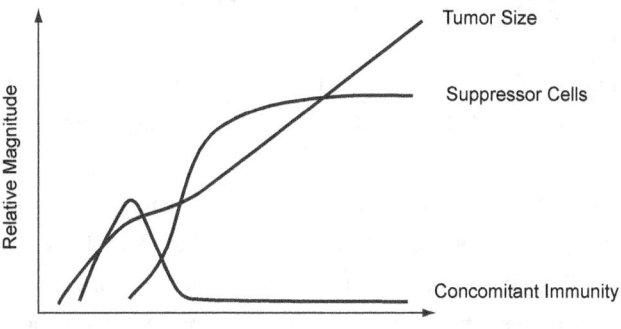

Fig. 49. North's view of the relationship through time between the expression of concomitant immunity, tumor burden and the generation of CD4 T suppressor cells

Around 1998, I decided that it would be of great interest to test some ideas on vaccination against and treatment of cancer. I summarize in point form several significant studies by others that inspired our investigations and the considerations that led to our experiments. It is easiest to understand why we undertook these studies if I state upfront the hypothesis they were designed to test. We proposed that most tumors are contained by a strong CTL response, and that most escape this response, most of the time, resulting in progressive tumor growth, by stimulating the generation of a substantial Th2 component of the immune response. These Th2 cells down-regulate the generation/activity of protective CTL. We refer to this as the *Th2-Skewing Hypothesis of Tumor Escape*. It is natural for the unknowledgeable reader to be somewhat surprised at this point. Have we not seen, as outlined in chapter 4, that this was the prevalent view in the late 1960s?[67, 68] Indeed it was. However, to give an idea of the contemporary scene, I need only point out that in four recent reviews on tumor immunology,[178-181] in all of which various and diverse mechanisms of tumor escape are listed, the Th2-Skewing Hypothesis is only mentioned in one, and then only as one of many.[178] This state of affairs is, I believe, an example of where previous insights have been lost in contemporary understanding.

A determination of the immune correlates of protection/progression seems to me to be the necessary starting point for the rational design of strategies of vaccination against and of immunotherapy of tumors. We need to know what kind of immune response is protective, and what other kinds of immune response are associated with tumor progression. I list below the main considerations that led to our Th2-Skewing Hypothesis.[10, 182]

1. North's studies seem parallel to those of many others, including Salvin, according to which the immune response to most antigens consists first of an exclusive cell-mediated phase, followed by the production of antibody and

the down-regulation of the cell-mediated response, compare Figure 49 with Figure 8.

2. In the mid-1970s, we had shown that this down-regulation of cell-mediated immunity was mediated by CD4 T cells,[78,79] subsequently identified as Th2 cells.

3. The most prominent immune cells, protective against the majority of tumors not bearing class II MHC molecules, appear to be CTL. Antibody is rarely protective.[67, 68] Thus the appearance of protective, concomitant immunity, and its subsequent disappearance, is consistent with the generation of non-protective antibody as the CTL response is suppressed.

4. During the last twenty years, investigators have found both in cancer patients, and in animals bearing tumors, that there are CTL[8] and antibodies[7] that recognize tumor-associated antigens. These CTL and antibodies are less prevalent in non-cancerous individuals and animals. The development of strategies to define these antigens, recognized by CTL and antibodies, and the genes encoding them, has been critical. Tumors of the same type often express some of the same tumor-associated antigens. These studies have brought a new level of analysis to the field. They show, in an unequivocal way, that the immune system recognizes and can naturally respond against cancers in humans and transplantable tumors in animals.

5. I think it important to convey the tone of the contemporary and prevalent view in the field of tumor immunology, and why I have a different slant on the subject. The sophistication of the immune system's response to antigens in general, and to tumors in particular, has become apparent, in exerting multiple modes of attack against foreign invaders.[178-181] These include attacks by the families of NK cells and γδ T cells. Many mechanisms of tumor escape against these diverse mechanisms have been documented, and it seems to me the prevalent view is that there are many mechanisms by which tumors can avoid one aspect or another of the immune system. I believe this is likely true. However, if there are many, roughly equally prevalent mechanisms of tumor escape, each allowing the tumor to grow progressively, it would make the realization of a generally effective vaccination strategy against cancer most problematical, if not impossible. The same would seem to be true of developing effective immunotherapy against cancer in animals or people, once the cancer is already established.

6. There are other reasons, besides the prevalence of concomitant immunity, for believing there is a commonality in the nature of the immune response against transplantable tumors. First, it is important to note that Robert North's studies on concomitant immunity, and on its regulation, involved several different tumor systems, reflecting different types of cancer.[183] In addition, tumor

immunologists are sometimes lucky enough to find a means of causing a normally lethal challenge of tumor to regress, and they then examine whether this manoeuvre is effective in other tumor systems. For example, North found that the same dose of whole body irradiation six days after implantation of radiation-resistant tumors caused regression in three of five tumor systems he examined, without exploring whether changing the radiation dose might reveal success with the "resistant" tumors.[183] A more recent study found that depletion of CD25+ T cells, before or around the time of tumor implantation, led to tumor regression of a challenge that was otherwise progressive, in *six out of eight* tumor systems examined.[184] The CD25 molecule is expressed on all activated T cells and is found on T_{reg} cells. I do not wish to digress here to discuss the grounds for believing the mechanism underlying this effectiveness of this manoeuvre is consistent with the Th2 Skewing Hypothesis of Tumor Escape. Rather, I wish to bring to the attention of the reader the effectiveness of this manoeuvre in a substantial majority of tumor systems studied. This finding again attests to a commonality in the nature of the regulation of immunity against these diverse transplantable murine tumors.

Our studies in tumor immunology: validating the Th2-Skewing Hypothesis of tumor escape in mouse models of human cancer

We chose to employ two transplantable murine tumors, among those most prevalently employed by tumor immunologists, to test the Th2-Skewing Hypothesis. Robert North had employed both in his studies. I should state at the outset that there was and is the widely held view in the literature that in both these cases tumor-specific CTL are important in resistance against the tumors. However, I think it fair to state that there was no clear idea as to what the immunological correlates of tumor progression might be, except for the observations made by North. Our findings support the Th2-Skewing Hypothesis in both tumor systems. Duane Hamilton, a graduate student, carried out most of these studies. I had suggested to Duane that he work on a particular project, but it was not going anywhere. Guojian Wei had begun to establish the tumor systems in our laboratory, based in part on some older studies by Calliopi Havele, my wife. Duane came in one day and requested that he completely change his field of study to that of tumor immunology. I immediately agreed.

Our evidence in favor of the Th2-Skewing Hypothesis is two-fold. First, Duane and I decided to inject mice with a number of lymphoma cells that led, in about half the subjects, to a rejection of the tumor, and in the other half to progressive tumor growth. We assessed the immune response at day thirty after tumor challenge, a time at which it was apparent in which mice the tumor was progressing and in which the tumor was contained. We found that tumor regression correlated with the generation of predominant Th1 responses, and tumor progression with mixed Th1/Th2 or predominant Th2 responses, see Figure 50. Second, we injected mice with low numbers of P815 tumor cells, well below the number resulting in progressive growth. We then challenged them some weeks later, as well as naive mice, with what is a normally lethal challenge. We found that those mice pretreated by injection with a low number of tumor cells resisted the tumor, whereas the naïve mice suffered progressive tumor growth.

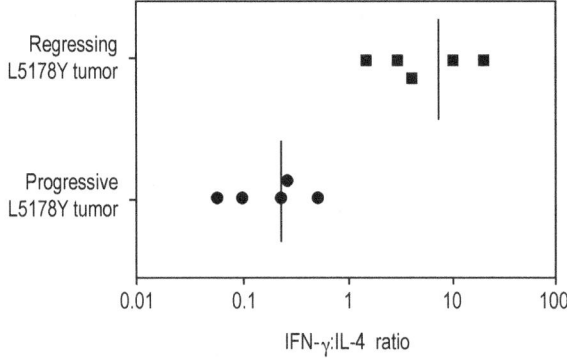

Fig. 50. Ratio of the number of tumor specific IFN-γ- to the number of IL-4-producing CD4 T cells in mice resisting and mice succumbing to a challenge of the lymphoma L5178Y that variably resulted in tumor progression or regression. Adapted from reference 185

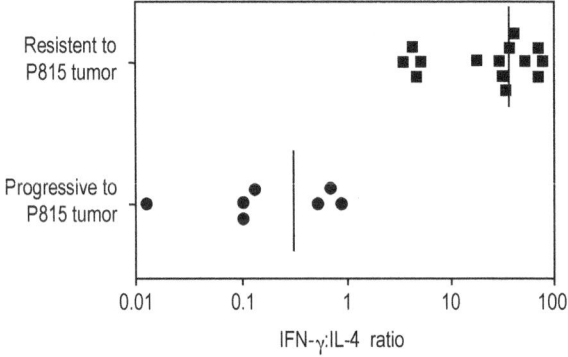

Fig. 51. The ratio of tumor-specific IFN-γ- to IL-4-producing CD4 T cells in naive mice succumbing and in vaccinated mice resisting a normally lethal challenge of the P815 mastocytoma. Resistant mice had been pre-exposed to a low number of tumor cells. Adapted from reference 185

Resistance to the tumor correlated again with a predominant Th1 response, and progressive disease with a mixed Th1/Th2 or predominant Th2 response, see Figure 51. These observations critically test the Th2-Skewing Hypothesis of Tumor escape[185] in the context of animal models employing transplantable tumors, and completely parallel our leishmania work. They demonstrate in this model the efficacy of the Low Dose Vaccination Strategy in the context of tumor immunology. We hope there will be opportunities to assess whether similar immune correlates of tumor regression and progression, as we have found in these two murine tumor systems, occur in other murine models and, more importantly, in human cancer.

Synopsis of Chapter 7

In the late 1980s, we began to apply our basic understanding of immune regulation to develop strategies of immunological intervention that would be of medical interest.

Clinical and animal studies show that strong, predominant Th1 responses correlate with protection against various intracellular pathogens capable of causing chronic disease, whereas chronic or progressive disease is associated with an immune response that has a substantial Th2 component. These immune correlates of protection/disease appear to hold in leprosy, tuberculosis, AIDS, and the leishmanisases. The murine model of cutaneous leishmaniasis is the most extensively studied animal model of these human diseases. Different strains of mice, infected with a standard challenge of a million parasites, generate either a stable Th1 response that contains the infection and are hence known as resistant, or they in time develop a predominant Th2 anti-parasite immune response associated with parasite growth and spread throughout the body, in which case they are designated as susceptible. We employed this murine model to explore strategies of both vaccination to prevent disease and strategies of immunotherapy to cure the patient.

We infected susceptible BALB/c mice with low numbers of parasites and hoped to thereby generate Th1 responses and Th1-imprints, rendering the mice resistant to a challenge of a million parasites that causes progressive disease in naïve mice. This low dose vaccination strategy worked spectacularly well. Our further studies were directed at developing a vaccination strategy that would be universally efficacious in a genetically diverse population. These studies led us to suggest that infection of diverse people or animals with a sufficiently low number of slowly growing organisms, such as BCG, would in time generate in all individuals an exclusive Th1 response, and so a Th1-imprint. All individuals would then resist a normally pathogenic challenge. We refer to this as the ultra low dose BCG vaccination strategy to provide universally efficacious protection against tuberculosis.

Certain manoeuvres, close to the time of infection with a million parasites, can result in normally susceptible BALB/c mice resisting the infection. This resistance is associated with a modulation of the long-term response from a Th2 to a Th1 mode. One means is by partially depleting CD4 T cells. A second manoeuvre is the administration of anti-IL-4 neutralising antibody. The IL-4 lymphokine is made by Th2 cells and stimulates their mul-tiplication, and is therefore centrally involved in the expansion of Th2 cells. Blocking such expansion allows Th1 cells to predominate.

These manoeuvres are ineffective in curing established infections in the standard model. However, infection of susceptible BALB/c with a million parasites rapidly results in a predominant Th2 response uncharacteristic of most human disease. We infected BALB/c mice with 10^4 parasites, an infection that leads to a substantial and chronic lesion asso-ciated with a stable, mixed Th1/Th2 response. We think this model better reflects most human disease. Such mice can be cured by giving them a judicious amount of monoclonal antibody that partially depletes CD4 T cells, or by giving them neutralizing anti-IL4 antibody. The large lesions of mice given these treatments resolve in a dramatic fashion, and their response to L major is modulated from a mixed Th1/Th2 to a predominant Th1 phenotype. Cured mice resist a challenge with a million parasites. We suggest these findings imply that appropriate and transitory depletion of CD4 cells works not only at the time of infection, but in established infections, to modulate the Th1/Th2 phenotype of a response towards a Th1 mode.

People infected with Leishmania donovani either resist the parasite or get ill with a syndrome called visceral leishmaniasis. They die unless treated with toxic drugs that kill the parasite. Most interestingly, those treated patients rarely get ill a second time, even though living in an endemic area. We examined the relative levels of IgG_1 and IgG_2 anti-parasite antibodies present in healthy, infected individuals and in patients before and after treatment. We found the IgG_1/IgG_2 ratio to be low in healthy infected individuals, indicating a predominant Th1 response, to be moderate in patients before treatment, indi-cating a mixed Th1/Th2 response, and to be low in treated patients, again indicating a predominant Th1 response. These observations are significant for two reasons. They show that the normal progression of a response from a Th1 to a mixed Th1/Th2 phenotype can be reversed in humans. This has implications for the treatment of some tuberculosis and of AIDS patients early in their disease. Second, there seems little doubt that this backwards modulation of the immune response is achieved by decreasing the load of parasite antigen. We suggest these findings mean that lowering the antigen load can lead, not only at the time of infection but in established infections, to the modulation of the Th1/Th2 pheno-type of an on-going, mixed Th1/Th2 response towards a Th1 mode.

Many studies in people with cancer, and in animals bearing tumors, show that these individuals have antibodies and CTL that can preferentially recognize tumor cells, and that such antibodies and CTL are more prevalent in these individuals than in their normal counterparts. These facts appear to unambiguously demonstrate that there is immune sur-veillance against cancer.

I argue that a broad review of the older and the contemporary literature supports the idea that there is a commonality in the regulation of the immune response in diverse tumor systems established in mice, and that there is a prevalent mechanism by which

most of these tumors escape from immune surveillance most of the time. Diverse observations led us to entertain the hypothesis that progressive tumor growth occurs when a substantial Th2 component of the anti-tumor response becomes sufficiently strong that it inhibits the anti-tumor CTL response. We refer to this as the Th2-Skewing Hypothesis of Tumor Escape. We tested this hypothesis in two highly studied murine tumor systems. In neither of these systems were the immunological correlates of tumor progression clearly established. One tumor was a mastocytoma and the other a lymphoma. We found that the Th2-Skewing Hypothesis of tumor escape held in both systems. Observations lead us to suggest this hypothesis may also be pertinent in the context of human cancer.

Chapter 8

PROSPECTS FOR DEVELOPING STRATEGIES OF PREVENTION AND INTERVENTION IN MEDICINE

Preface

In the preface to this book, I indicated my belief that the virtually exclusive emphasis on observations and considerations at the cellular/molecular level is detrimental to progress in the field. There is also considerable pressure to work within current frameworks if one is to be competitive in gaining research funding. I believe that some central and current frameworks are misleading. It may be partly for this reason that I have been unable to obtain major funding over the last ten years.

My research has been supported over many years by the people of Canada, through grants from the Medical Research Council, and from the more recently established Canadian Institutes for Heath Research. I am deeply grateful for this support. To be given such trust is indeed a privilege. It is important that the distribution of public moneys is carefully administered and that the process of adjudicating money for research is as objective as possible.

Securing funding for research is a major portion of a scientist's work, and funds are not always available or awarded for projects that might otherwise show promise. In this section, I will review some of my own work and ideas that have not received funding, and so have not been experimentally explored.

Introduction: the basis for our considerations

No one can provide an incontrovertible basis for a discussion of potential strategies of intervention. I list the central propositions I think are worthy of exploration, formulated from observations in the literature and on the basis of our studies on the regulation of the immune response. I then explore how one might employ these propositions to envisage strategies of immunological prevention, or intervention, to prevent or relieve disease. I give my slant on the possibilities in four of the five pertinent areas of medicine: allergies, autoimmunity, cancer, and infectious diseases—leaving out transplantation. It should go without saying that these speculative proposals for exploration are meant to be illustrative rather than comprehensive. I start by giving a list of propositions on which my speculations are based, and indicate the grounds for favoring them.

(i) Proposition concerning the activation and inactivation of CD4 T cells

The cooperation of CD4 T cells is required to optimally activate naive CD4 T cells, and a single or very few CD4 T cells, are inactivated on interacting with sufficient antigen over a considerable time. The theoretical grounds and indirect experimental support for this proposition have already been described. However, decisive and direct evidence supporting it is not available at the time of writing.

(ii) States of immune deviation exist

Basic studies show that different states of immune deviation can exist, for example Th1, cell-mediated immune deviation, and Th2, humoral immune deviation. We also anticipate, though it is not so well studied experimentally, the existence of dominant anti-inflammatory immune states associated with T_{reg}-like cells, and the predominant production in humans of IgG_4 and IgA antibodies. All these antigen-specific immune states appear to be associated with antigen-specific T cells that have a capacity for inhibiting the generation of opposing classes of immunity.

(iii) Those antigen specific T cells that regulate the generation of immune responses normally act through the operational recognition of linked epitopes, independently of whether they facilitate or inhibit responses.

We have seen that this is true, for example, of the CD4 T cells able to facilitate the activation of B cells, and of those CD4 T cells, presumably Th2-like cells, that are generated during the course of a vigorous humoral response and that act to inhibit the generation of DTH. It seems that only if T cells act by such means, to regulate the fate and activity of other lymphocytes, can critically important physiological features of regulation be achieved. For example, only in this case can one envisage how non-interference of the inactivation of CD4 T cells by peripheral self antigens, by the activity of Th cells involved in concurrent responses to foreign antigens, be achieved. Only in this case can one envisage how the Th1/Th2 phenotype of concurrent responses, to two non-crossreacting antigens, both responses being generated in the same lymphoid organ, be independently determined. I should add that most immunologists have inferred from the evidence that T_{reg} cells specific for one antigen act by being triggered in an antigen-specific fashion to release long-range molecules able to inhibit responses to other, non-crossreacting antigens. I find this inference difficult to accept, given the physiological implications of such a mechanism. The observations on which this view of the mode of action of T_{reg} cells is based come from somewhat artificial systems and so may be misleading. I am relieved to find that there is evidence that human T_{reg}-like cells act by the operational recognition of linked epitopes in their role of down-regulating the production of cyokines by Th2 cells. These observations were made in a much more natural system.[124]

(iv) States of immune deviation are not immutable

The recognition that immune responses to an antigen can often be operationally biased towards, or locked into, one mode or another is fundamental. However, if we understand the basis of such locks, we might be able to pick them. We consider here some ideas and observations that might help us to do so.

It is important that the formation of locks is regulated so that they do not become overwhelmingly strong. For example, consider a state of humoral immune deviation to an antigen A. We would expect, if the TsDTH cells associated with such a state were generated in a great excess of what is required to suppress DTH responses to A, that these A-specific TsDTH cells would suppress primary DTH responses to other antigens that crossreact only slightly with A. In this case, the immune system would have lost its flexibility to mount an appropriate response. It is therefore plausible that the generation of regulatory cells is itself so regulated that that they exist at a level not much greater than the level needed to fulfill their primary role. There is an instructive and amusing *gedanken experiment* that illustrates that locks can hardly be irreversible if T cells act by the operational recognition of linked epitopes. Consider an animal in which the response to the antigen H has been locked into a humoral mode, and in which the response to another antigen D has been locked into a cell-mediated, DTH mode. What happens when we challenge such an animal with the conjugate H-D? Presumably, a tug of war ensues.

This consideration sets the scene for examining an enlightening and elegant study. Lymphocytes were obtained from two classes of individuals living in the same geographical area. Some were allergic to a prevalent allergen, denoted as A, and others were non-allergic. It was possible to isolate, by a beautiful technique, A-specific Th1, Th2, and T_{reg} cells from the lymphocytes harvested from both kinds of individual, and to enumerate their frequency. The bulk properties of lymphocytes from patients and non-allergic individuals characteristically differed from one another in that, for example, the lymphocytes from allergic individuals could be readily stimulated by A to produce substantial amounts of Th2-type cytokines, whereas lymphocytes from non-allergic individuals could not. Although lymphocytes of patients and non-allergic individuals contained all three types of antigen-specific T cells, their relative frequencies were different. Non-symptomatics had relatively more T_{reg} cells compared to Th2 cells than allergic patients. Moreover, it was possible to do reconstruction experiments with purified and isolated T cells, so that the reconstructed population of lymphocytes from a non-allergic individual had the frequency of the different T cell types found among the lymphocytes of allergic individuals. Now the population of reconstructed lymphocytes behaved similarly to an intact population of lymphocytes from an allergic individual and differently from the parental population from which they were derived.[124] This type of experiment convincingly demonstrates that the number and relative proportion of the different T cell types determines the behavior of the population as a whole. Consider the situation where the immune response to an antigen C is in a predominant mode M1, and we wish to modulate the response towards mode M2. We may be able to first establish a state of immune deviation to an unrelated antigen D into the mode M2, and achieve modulation of the immune response to C, towards M2, by immunizing with the conjugate C-D. Perhaps immunization with both C-D and D would be more efficient as the D-specific T cells would be less subject to regulation by C-specific T cells, thus giving the D-specific T cells the edge, in the envisaged tug of war.

(v) Implications of the Threshold Hypothesis and Macroimmunology

We have seen repeatedly from our basic studies that a combination of the number of CD4 T cells and the amount of antigen determines the Th1/Th2 phenotype of the immune response, as anticipated on The Threshold Hypothesis. Consider a nominal antigen Q that can be processed into peptides q1, q2, q3, ... qn, that can bind to host class II MHC molecules. Consider also what happens, according to the Threshold Hypothesis, when Q impinges upon the immune system. The events surrounding the activation of q2-specific CD4 T cells depends on the number of CD4 T cells specific for other q peptides and the amount of Q administered. If cooperation is weak, a Th1 response will be generated; if strong, the result will be the generation of Th2 cells. The fate of q-specific CD4 T cells is determined, according to this view, by collective rather than individual circumstances. We refer to this collective fate as *coherence* to reflect the fact that the determination of the Th1/Th2 phenotype of effector CD4 T cells, arising from the activation of diverse Q-specific CD4 T cells, is interdependent and so coherent. The fact that there is a collective fate of Q-specific CD4 T cells allows one to easily control the Th1/Th2 phenotype of all the CD4 T cells associated with the immune response by adjusting various parameters of immunization, for example the dose of antigen or the number of responding CD4 T cells. This framework contrasts with others, such as the Two Cell Framework. We would like to emphasize the importance of the ability to control the Th1/Th2 nature of an immune response through adjusting broad parameters of immunization. The recognition that broad principles govern the collective fate of the diverse CD4 T cells specific for the same nominal antigen, and the recognition of the existence and use of simple rules governing these fates, provides the foundation for a *macroimmunological* approach to immunological intervention.[182]

(vi) The decision criterion that determines the Th1/Th2 phenotype of on-going immune responses

Knowledge of the relationship, if any, between the decision criteria controlling the Th1/Th2 phenotype of a primary and of an on-going immune response must come from observation. It is not *a priori* clear that there is a relationship. Our observations in mice, discussed above, show that the temporary and partial *in vivo* depletion of CD4 T cells during a chronic immune response to *Leishmania major* can modulate a mixed Th1/Th2 response to a Th1 mode. Our studies in patients with visceral leishmaniasis, where the administration to patients of anti-parasitic drugs results in cure and involves a similar modulation of the Th1/Th2 phenotype of the anti-parasite response, lead us to suggest that reduction in antigen load can have a similar affect on on-going responses as on primary responses. These observations support the proposition that the Th1/Th2 phenotype of on-going immune responses can be modulated towards a Th1 mode by a judicious and temporary reduction in the number of CD4 T cells and/or a reduction in antigen load. These are important concepts, if valid, as they would then provide the basis for modulating on-going immune responses in a desirable direction. We shall consider such a possibility in the early stages of AIDS, shortly after the infected individual has seroconverted, and in the context of cancer immunotherapy.

Allergies

I have outlined above some recent and exciting developments in the field of allergy. It appears that non-allergic individuals, living in a geographical area where allergy is prevalent, usually express a state of anti-inflammatory immunity against the allergen. Specific immunotherapy, by administering antigen to the allergic patient, is problematic, as it may trigger acute inflammation and anaphylaxis, potentially leading to death. Immunotherapy, when achievable, drives the immune response towards an anti-inflammatory immune state. We consider any strategy that might be effective in improving the practicality of specific immunotherapy should attempt to avoid the triggering of acute inflammation in allergic individuals. We consider a couple of possibilities.

Acute inflammation occurs in allergic people when antigen cross-links IgE antibody bound to mast cells. This cross-linking leads to the rapid release by the mast cells of preformed mediators of acute inflammation, such as histamine. We discuss the basis of two approaches that try to make immunotherapy more realistic by avoiding the triggering of acute anaphylaxis.

Both strategies are based upon the fact that B cells and antibody recognize antigen in a manner different from CD4 T cells. It therefore might be possible to find a form of antigen able to stimulate T cells, which does not readily cause mast cell degranulation due to the existence of mast cell-bound IgE antibody specific for the allergen.

We have seen that primed CD4 T cells specific for a peptide p bound to a class II MHC restriction element, R, can be stimulated to produce cytokines by the presence of p and of APC bearing R. It is believed that about half the resident peptides, bound to class II MHC molecules on the surface of an APC, come off the class II MHC molecules in about twenty-four hours, allowing the peptide p, if present, to bind to the empty grooves of these class II MHC molecules. We refer to this process as *peptide decoration of the APC from the outside*, to distinguish it from the normal intracellular process by which peptides are loaded onto class II MHC molecules.

Antibodies bind well to the intact protein they are raised against, and poorly to a protease digest of the nominal antigen. However, as we have seen, such digests can stimulate CD4 T cells. One approach to the development of specific immunotherapy is to use such a mixture of peptides, derived from the antigen against which the individual is allergic, resulting in CD4 T cell stimulation without substantially triggering the interaction of antigen with mast cell-bound IgE antibody, and so avoiding anaphylactic reactions.[186] It appears that such administration of peptides inactivates the allergen-specific of Th cells or deviates them into another mode.

Another approach we envisage also attempts to exploit the fact that antibodies and T cell receptors bind to different forms of the antigen. In general, small class II MHC-binding peptides are not very immunogenic. This is presumably because there are not many CD4 T cells specific for just this one peptide, precluding effective CD4 T cell cooperation essential for the sustained expansion and activation of these CD4 T cells. Consider an individual allergic to an antigen A, whose mast cells are loaded with anti-A IgE antibody. Is there a way we could design an antigen that is expected to be highly immunogenic in stimulating CD4 T cells, would not interact well with anti-A IgE antibody, and yet would interact with CD4 T cells specific for the nominal antigen A? If there was, we might be able to employ such an antigen to stimulate the CD4 T cells, specific for A, from a Th2 to T_{reg} mode, whilst minimizing the risk of causing anaphylaxis, as the antigen would be designed to react poorly with A-specific IgE antibody.

We know that the stimulation of CD4 T cells specific for the nominal antigen A requires it to be processed into a series of peptides, of the order of 10-20 amino acids long, say a_1, a_2, a_3, ... a_n. Some of these "a" peptides will bind to the grooves of the host's class II MHC molecules, and so be presented and stimulate their corresponding CD4 T cells. Consider the gene for A; we can divide it into a series of segments that code for oligopeptides about fifteen amino acids in length. We can synthesize a gene where these segments are in a different order than they are in the gene coding for A. We could design this gene so that the encoded protein would, when processed, give rise to the same a_1, a_2, ... a_n peptides as results from the processing of A, as well as some more. My graduate student David Kroeger suggested, as we discussed this idea, that we call this synthetic protein an "isopeptidogen" of A, as it generates the same peptides as A when processed. We anticipate the isopeptidogen of A will react poorly with anti-A IgE antibody, as its three dimensional structure will be very different, and so administration of the isopeptidogen may be useful for desensitizing individuals allergic to A, whilst minimizing the likelihood of anaphylactic reactions. In fact, the isopeptidogen would most likely not form a well-defined three-dimensional structure, and would be insoluble. It may therefore be possible to give it locally, in a manner that it could be a source of antigen to stimulate immune cells. A further advantage of such administration would be that antigen, primarily limited to a local site, would not so readily cause anaphylaxis.

Autoimmunity

The proposals that the sustained interaction of a single CD4 T cell with antigen presented by APC results in its inactivation, whereas its sustained activation requires its interaction with other CD4 T cells, mediated by the recognition of APC-presented antigen, should allow one to develop powerful ways of inducing tolerance. We consider the implications of these proposals in the context of the mouse model of autoimmune diabetes. These mice, known as NOD (non-obese diabetic) mice, represent a remarkably good model of human autoimmune diabetes.[4, 187, 188]

We described in Chapter 6 how studies in NOD mice and multiple sclerosis reveal the phenomenon of epitope-spreading,[104, 105] which is anticipated if there is a requirement for CD4 T cell cooperation in the activation of CD4 T cells. Moreover, blocking the generation of activated CD4 T cells specific for the first antigen recognized also blocks the generation of activated CD4 T cells to the other antigens that are normally seen as autoimmunity develops, and so prevents diabetes.[104]

We envisage that the cooperation between different CD4 T cells is mediated through their recognition of peptides presented by the same APC, most probably a B cell. It is known that expression of antigen systemically by all dendritic cells, or the systemic administration of large amounts of a class II MHC-binding peptide that presumably leads to the systemic "decoration" of APC, results in the inactivation of the corresponding CD4 T cells. It would appear that in such cases there will be many more APC presenting the antigen than there are CD4 T cells specific for the antigen—a situation unfavorable for CD4 T cell cooperation. Interestingly, such situations appear not only to result in the inactivation of naïve CD4 T cells but also of activated CD4 T cells, and so may be applicable not only in preventing but in reversing autoimmunity. [189, 190]

Another approach to relieve autoimmunity in an antigen-specific fashion may be realistic. We illustrate the idea by a hypothetical example. In multiple sclerosis, it is believed an inflammatory, cell-mediated response is directed at the insulating myelin sheath of nerve cells. The inflammatory response causes damage. We are not concerned here how multiple sclerosis arises.

In the chronic, relapsing form of the disease, there are severe episodes interspersed with quiescent periods. One hypothetical explanation for this pattern is based on what we known about how on-going immune responses are regulated. We have seen that the dose of antigen is important not only in determining the Th1/Th2 phenotype of primary immune responses, but of on-going immune responses. One potential way of accounting for the chronic, relapsing form of the disease is that a severe episode is associated with cell-mediated inflammation around the myelin sheath, caused by a predominant Th1 response and resulting in substantial release of myelin antigen. As the antigen load increases, the immune response is consequently modulated in time from a predominant Th1 into a mixed Th1/Th2 mode. There is consequently less damage and, as this occurs, a quiescent period ensues. As a result of this quiescence, the amount of myelin antigen released decreases, and so the response is again modulated in time from a mixed Th1/Th2 to a more predominant Th1 mode, in the same manner as we described above when patients, with visceral leishmaniasis, are treated with anti-parasite drugs.

If this hypothetical explanation for how the disease can wax and wane is valid, the ratio of the IgG_1 to IgG_2 isotypes present in IgG anti-myelin antibodies present in patients should vary accordingly. The ratio should be small during attacks, and an increase in its size should correlate with the appearance of a quiescent phase. If this prediction were verified, it would confirm the supposition that severe episodes and quiescent periods are associated with different classes of immunity to myelin antigens. In this case, it might be ethical to explore the possibility of administering myelin antigens conjugated to other antigens against which there is a strong Th2 response, in order to establish an immune state against myelin antigens associated with quiescent periods.

Infectious Diseases

Effective vaccination of people in the West has only been achieved, in a manner publically and widely acknowledged, against pathogens that cause acute disease and are contained by a vigorous antibody response. Effective vaccination against other pathogens, particularly intracellular pathogens with the potential for causing chronic disease, and that are best contained by a cell-mediated response has not been formally accomplished. I feel I should note, however, that some people in the Middle East have been deliberately exposed to sandflies infected with *Leishmania major* to protect them against cutaneous leishmaniasis, a process called ***leishmanization***.[191] It seems likely that leishmanization is a folk practice that reflects a process of natural infection with low numbers of parasites.

I start by considering vaccination against HIV, the virus that causes AIDS, as an illustrative example. I shall also consider whether and how it might be possible to reverse AIDS progression at the early stages of the disease, shortly after seroconversion. I shall illustrate further considerations in discussing the possibilities of achieving effective vaccination against and treatment of tuberculosis.

Vaccination against AIDS

Before considering what principles might guide the development of an effective strategy of vaccination against AIDS, I would like to indicate why I believe a consideration of such principles is critical. It is argued by some that one should explore diverse possibilities without prejudice. Consider what such a study might look like. There are proponents for different forms of the vaccinating agent, such as purified components of the virus, whole but dead virus, various vectors that carry HIV genes that have limited ability to replicate, and other vectors that survive much longer. It is not clear what dose of these different alternative vaccination agents would be optimal, nor the site of inoculation of the putative vaccine, nor whether to employ ***adjuvants***, substances given with a vaccinating agent to increase its immunogenicity. Once one has vaccinated the animal, how should one challenge the subject to assess protection? Should one challenge with a barely pathogenic strain or one that causes rapid disease, or perhaps one should employ a challenge of intermediate virulence? Should one use multiple animal models, or just one and, if so, on what basis should this one be chosen? How long after vaccination should the challenge be? Suppose you want to carry out the most encompassing study. You will need to examine five different forms of the vaccinating agent, say, at three different sites of primary inoculation, exploring four different doses of the agent, in four animal systems, with three challenges of different strength, with five different adjuvants, and assessing the protection achieved at three different times after the initial vaccination. Such an encompassing study would require 5 x 3 x 4 x 4 x 3 x 5 x 3 = 7200 different groups. The numbers I have chosen can of course be argued with. Nevertheless, I believe this illustrates the dilemma of the all-encompassing approach. Moreover, it is unclear whether this strategy will uncover the critical, because effective protocol, of vaccination.

Consider our actual experience in realizing the low dose vaccination strategy in the mouse model of cutaneous leishmaniasis. We started with a conceptual framework, as previously outlined. With this framework in mind, we injected nine groups of mice belonging to a "susceptible" strain with different numbers of parasites, but by the same route and without adjuvant. We then followed the development or lack thereof of parasitemia, and the type of immunity generated. Mice infected with few parasites did not develop high levels of parasitemia and produced a predominant Th1 response, whereas mice infected with larger numbers of parasites developed substantial parasitemia and predominant Th2 responses. These observations were anticipated within our framework. We further anticipated that infection with few parasites would generate not only predominant Th1 responses but also in time Th1 imprints, and so protection against a challenge with a high number of parasites that causes progressive parasitemia in naive mice.

We first challenged the mice infected with a few parasites one month after this first infection. We observed little if any protection. Believing our approach still worthy of further exploration, we repeated the experiment a second time, but waited this time for two months after the infection with a few parasites before the challenge. The mice were dramatically resistant. At this stage, we have one foot in the door. It is now easy to determine whether giving the parasite by, say, different routes and with different adjuvants improves the protection in any way, because we have a positive standard of protection against which we can assess any potential improvements. In practice, we employed only thirteen groups of mice to establish this protocol for achieving protection. We injected nine groups of mice with different numbers of parasites in the first experiment, and then set up two further groups, twice, to see whether infection with low numbers generated resistance upon challenge. A clear and valid hypothesis as to what type of immunological response is protective, and a knowledge of how this class of immunity is regulated, allowed us to accomplish this goal. Naturally, the success of such an approach will only be realized if the framework employed is substantially correct. It seems to me that the identification of the immunological parameters, discriminating between the responses of healthy infected individuals, and of patients, is the crucial starting point for vaccine design. These observations tell us what type of immunity we should favor by vaccination, and what type to avoid.

In looking at current views, it appears that there are two main, clear, but contradictory positions as to how effective vaccination against HIV-1 might be achieved. Most suggest that successful vaccination should guarantee upon natural infection both cell-mediated immunity, in the form of CTL, and antibody, particularly in the form of ***neutralizing antibody***.[192] The neutralizing antibody would bind to newly synthesized virus released from infected cells, leading to removal of the virus and preventing the infection of new target cells. It should be noted that neutralizing antibody can be generated, but that HIV-1 has a means of generating variants quickly and in a manner that most neutralizing antibody, effective against the variant that induces the antibody, is usually ineffective against other variants that become prominent.[193]

The contrasting view is that vaccination must aim to guarantee a predominant Th1 response associated with potent CTL generation upon infection by HIV-1. I share this view for three reasons.[194, 195] First, there are highly exposed and infected individuals who produce predominant and stable Th1, CTL responses, and who are symptom free. For example, some female sex workers in Nairobi belong to this group.[5,6] What is additionally remarkable about them is that they do not seroconvert, and remain healthy over many years, though presumably being exposed on different occasions to major viral variants, or clades of the virus, through their sexual work. Second, all individuals infected with HIV-1 are relatively healthy during their "honeymoon period", when their immune system initially mounts a predominant Th1, CTL response, and for a period after they seroconvert, before they start going through the progressive stages of disease. Both these sets of observations strongly support the idea that a predominant Th1, CTL response is protective. Why then not aim for having the best of both worlds, as most suggest? It is an assumption, and one that I feel is highly questionable, that this is possible. We have seen repeatedly that distinct CD4 T cell subsets tend to inhibit each other's generation, and often each other's effector activities. It may well not be possible to have the best of both worlds, involving a mixed Th1/Th2 response with optimal CTL immunity and the optimal production of neutralizing antibody. I think it reasonable to conclude, from observations made in the natural situations described above, in which effective resistance against the virus is clearly apparent, that predominant Th1, CTL responses are protective, and to employ this proposition as the starting point for imagining how effective vaccination might be achieved. [6,166]

Here I describe and outline the reasons underlying my preferred, but untested, strategy to provide universally effective vaccination against HIV-1. This strategy is based upon similar principles as the strategy, already described, to achieve universally efficacious vaccination against tuberculosis, namely the ultra-low dose, BCG vaccination strategy. I recall the essence of this strategy before outlining how additional considerations might lead to effective vaccination against AIDS.

I have proposed in Chapter 7 that administration of very low numbers of BCG can lead to resistance to tuberculosis in a genetically diverse population. The number of BCG would be chosen to be so low that it does not induce antibody in any individual, but in time induces Th1 responses and generates Th1 imprints in all.

I have suggested also that such Th1-imprints will constitute effective vaccination against tuberculosis.

An important feature of this scheme is that it allows the generation of a BCG-specific Th1 imprint in genetically diverse individuals. Critical to this important feature is the fact that BCG is a slowly replicating entity. I suggest for similar reasons that it will be most helpful to employ a slowly replicating entity to vaccinate against HIV-1 to achieve universally effective vaccination. In addition, it would appear unlikely that vaccination with an attenuated form of HIV-1 would be acceptable, as it is unlikely that virologists can sufficiently guarantee that the attenuated strain would under no circumstances revert to become virulent. However, BCG is the vaccine that has been used in humans worldwide more frequently than any other, with the most marginal of side effects.[21] Moreover, people have inserted HIV genes into BCG in a manner that allows them not only to be expressed to produce protein but the BCG vector has been employed to immunize animals and raise immunity against the protein coded for by the inserted gene. Three further considerations/observations make me enthusiastic that BCG vectors expressing HIV-1 genes might be ideal platforms for vaccinating against HIV-1.

First, HIV-1 has only a few genes, and it is possible to put single HIV-1 genes into different BCG vectors and immunize with a mixture of these vectors. This seems to be a safe way of immunizing against a series of HIV-1 antigens, employing entities that replicate, namely the BCG vectors. Second, a graduate student, Carl Power, examined in the early 2000s how the Th1/Th2 phenotype of the immune response against the protein, encoded by the vector and referred to as P_v, depended on the number of vector-carrying BCG employed for infection. Infection with low numbers of these BCG vectors resulted in a Th1 response to both P_v and BCG, whereas infection with a higher number resulted in a mixed Th1/Th2 immune response to both P_v and BCG. Thus, infecting with a low number of a mixture of BCG vectors is expected to result in a Th1 response to the HIV-1 encoded antigens.[196] Third, we anticipate that infection with ultra-low numbers of an appropriate mixture of BCG vectors will result in a Th1 response and a Th1-imprint against the HIV-1 antigens in all individuals infected. It seems to me this strategy is promising for achieving effective vaccination against AIDS and is worth testing. However, I have been unable to get funding to explore this strategy. I recently discovered Jon Cohen, in his book *Shots in the Dark* on the attempts to generate an effective vaccine against AIDS, quotes Jonas Salk as saying that I thought infection with low numbers of BCG vectors would constitute the perfect vaccine. Jonas shared my enthusiasm.[197]

At one time I had not only talked with Jonas Salk on roughly a weekly basis for over a year about how one might achieve effective vaccination against AIDS, but Jonas had given me a modest but significant amount of money from one of his foundations for AIDS research. This expression of faith in my ideas was of critical support later, after Jonas died, when I found it difficult to get funds to continue my research, and could use Jonas' remaining funds.

Immunotherapy of AIDS at early stages of the disease

Might it be possible to reverse AIDS progression shortly after seroconversion occurs and before the disease has become advanced? I should like to propose a strategy based upon a consideration of four sets of observations.

First, we have described above how mixed human Th1/Th2 responses against *Leishmania donovani* can be modulated backwards to a predominant Th1 mode in visceral leishmaniasis patients by administering anti-parasite drugs.[176] Such a modulation is thus in principle possible. I have argued above that such modulation most probably occurs due to a lowering of the parasite burden on treatment, and hence a lowering of the effective antigen load. I anticipate that a similar and appropriate lowering of the viral burden early in AIDS, shortly after serconversion, and when there is a significant but modest Th2 component to the immune response, could modulate the immune response of the infected individual to a predominant Th1 state. This is the state associated with the honeymoon period, and present in resistant sex workers. Such a modulation would thus reverse the progression of AIDS. Furthermore, the return to such a state, associated with a low viral load, would also make the individual less infectious to others.

Second, this line of reasoning is supported by an examination of the nature of the immune response in seropositive, HIV-1-infected individuals, whose disease does not progress or progresses very slowly. This examination shows that the IgG anti-HIV-1 antibodies of such individuals are predominantly of the IgG_2 isotype, associated with a predominant Th1 anti-HIV-1 response. This finding again leads to the conclusion that stable, predominant Th1 responses are associated with effective containment of HIV-1 and so protect against AIDS.[198]

Third, we examined in our laboratory in the mid-1980s whether it is possible to modulate immune responses of mice backwards from a humoral to a cell-mediated mode and, if so, to ascertain the conditions under which such modulation occurs.[199] We immunized mice in a manner to produce IgG antibody but not to express DTH to sheep red blood cells (SRBC), and then harvested their spleen cells and cultured them under a variety of conditions. In most cases, the antibody response decayed over a few days and potent, SRBC-specific DTH-mediating cells appeared. I now think that this modulation occurs because, when we put the spleen cells into culture, we are in effect diluting them compared to their natural *in vivo* state in the intact spleen, so CD4 T cell interactions become less robust, and the phenotype of the immune response drifts towards the Th1 pole. The modulation of the response towards cell-mediated immunity nearly always happened but the rapidity of this modulation depended upon the density at which the spleen cells were cultured and the amount of antigen present. Interestingly, there was one circumstance in which all immunity seemed to disappear. This occurred when we cultured the spleen cells without adding any additional antigen.[199] When the amount of antigen is very low, it appears that the antigen-mediated cell interactions required to sustain or generate immunity are too weak, and the antigen-specific lymphocytes peter out. This finding was dramatic in the sense of being very clear.

Fourthly, some of the findings on treating AIDS patients with anti-viral drugs strike me as highly interesting in view of the three sets of observations just described. HIV belongs to a group of viruses known as retroviruses, and drugs that act against them are called anti-retroviral drugs. Patients are sometimes given a course of treatment involving the administration of multiple anti-retroviral drugs. When this course is designed to dramatically reduce the retroviral load, it is referred to as *highly active anti-retroviral therapy (HAART)*. Sometimes there are sufficiently strong side effects of such treatment that the physician and patient decide to interrupt it. In this case, there is usually a rapid and dramatic increase in detectable virus, known as *viral rebound*,[200] indicating that protective immunity has most probably been undermined by HAART. In addition, it has been reported that in some individuals in which HAART has been interrupted a few times, so the immune system is exposed intermittently to more HIV virus, that further interruption of HAART therapy does not result in viral rebound.[200] It seems to me that all these observations on viral rebound, and the occasional lack of viral rebound after interrupted HAART therapy, might well be understood in terms of our third set of observations: the SRBC-specific lymphocytes among spleen cells immune to SRBC disappear when the spleen cells are cultured without deliberately added antigen, i.e. when very little if any antigen is present. If HAART results in a great reduction in viral load, it may well be that the virus is not present at a level where it can mediate the interactions between leucocytes required to efficiently generate/sustain immunity to the virus. Is there any way of overcoming this problem, if this conjecture is correct?

We anticipate that at the start of HAART treatment, and the consequent drop in viral load, that the patient's immune response will be modulated from a mixed Th1/Th2 to a predominant Th1 mode. Once this has occured, the prevalence of antibodies of IgG_2 isotype will be greater than antibodies of IgG_1 isotype, as seen in seropositive nonprogressors.[198] We anticipate that if HAART therapy is halted at this time, viral rebound will not take place as the patient's Th1 immunity, associated HIV-specific CTL, will contain the virus. Further HAART at this stage would further reduce the viral burden and undermine the sustained generation of the protrective CTL, Th1 response. Monitoring of the IgG isotypes among anti-HIV IgG antibodies can be used to ensure the response remains in a predominant Th1 mode and, if it starts evolving towards a Th2 mode, HAART therapy can be reinstated for a short while. This would appear to constitute a potential and personalized treatment of AIDS at early stages.

Vaccination against and treatment of tuberculosis

I proposed earlier that the universal establishment of a mycobacterium-specific Th1-imprint upon the immune system would provide universal protection against infection by *Mycobacterium tuberculosis*. This premise is not obviously valid. I first address why I think this premise is sufficiently plausible that it is worthy of exploration.

A sound approach to vaccination requires an understanding of why immunity succeeds in providing protection under some circumstances and fails under others. One way of trying to gain an insight into why these different outcomes occur is to identify the immunological parameters that discriminate the immune states associated with containment of the pathogen from those associated with chronic or progressive disease.

I used to sometimes joke with my colleagues that, if you want to find a quick way to depression, you should intensely study the literature on the immune correlates that distinguish between containment of *Mycobacterium tuberculosis* and those associated with disease. In doing so, I found myself uniquely frustrated, to put it lightly, over a number of years. When various parameters of the anti-mycobacterium immune state are compared between tuberculosis patients and healthy infected individuals that contain an infection of *Mycobacterium tuberculosis*, some clear but somewhat incomprehensible conclusions can be made.[201, 202] For example, the average level of anti-mycobacterium-specific IgG antibody is clearly higher in patients than in healthy contacts. These studies lead to the conclusion that *on average* there is a greater antibody component to the immune response in patients than in healthy infected.[201] My frustration arose because this tendency was only apparent when the immune parameters of many patients were compared with those of many healthy infected; however, the level of IgG antibody in many *individual* patients was actually considerably lower than in most healthy contacts! There appeared to be no way of interpreting these observations in a simple and plausible manner.

During her PhD studies with me, Juthica Menon developed the low dose vaccination strategy in the murine model of cutaneous leishmaniasis. She later tried, for a couple of years, to define immune parameters that distinguish the immune state of tuberculosis patients from those of healthy infected individuals. These studies were most frustrating and not making progress. Juthica assessed the relative levels of different isotypes of IgG antibody among the mycobacterium-specific antibody of patients and healthy contacts, including the levels of a doctor friend of mine, Andrew Judd, who had had many tuberculosis patients in his practice. It seemed to me his anti-mycobacterium immune state must represent a state of exceptional resistance. Looking at his immune parameters, I noticed that the level of IgG_2 antibody was much greater than the level of IgG_1 antibody, representing a predominant Th1 response. I suggested one day to Juthika that she calculate this IgG_1/IgG_2 ratio for the anti-mycobacterial antibody of all the patients and of all the healthy infected individuals she had studied.

Juthica showed me the results of the calculations the next day. The value of this ratio for healthy contacts varied from 0.001 to 0.3, and for tuberculosis patients from 0.001 to 100. I readily formed a hypothesis to explain these findings. These thoughts occurred so readily because I had traversed many paths over the last few years in my vain attempts to understand the difference in the nature of the immune response to *M tuberculosis* in patients and healthy infected individuals. It seemed self-evident that the qualitative nature of the immune response in tuberculosis patients with an IgG_1/IgG_2 ratio above 1 was different from that of healthy contacts. From what we already knew about the role of Th1 and Th2 cells in the production of different isotypes of IgG antibody, the immune response of these tuberculosis patients most likely had a greater Th2 component than that present in healthy contacts. We came to refer in our discussions to these patients as having **type 2 tuberculosis.** We envisaged that their anti-mycobacterial response had a significant and detrimental Th2 component.

What could be the cause of failure of the immune response in tuberculosis patients that had a similar IgG_1/IgG_2 ratio as healthy contacts? My previous struggles to understand the literature on tuberculosis led me to what seemed an obvious possibility. These individuals generate an immune response that is qualitatively similar to that generated in healthy contacts, but the response is too *weak* to contain the infection. In this case, the mycobacterial burden would increase and substantial lesions would form. We refer to this hypothetical form of tuberculosis as **type 1 tuberculosis.** Two considerations led me to find this possibility appealing.[171]

We had injected a million *Leishmania major* parasites subcutaneously into the footpad of mice belonging to resistant strains, and followed what happened. The immunological community had designated these mice as resistant as they contained the parasite, due to the development of a sustained and stable Th1 response. However, before control of parasitemia was achieved, over some weeks, a big lesion developed in the footpad of the mice, resulting roughly in a doubling of the foot's size! It takes considerable time to develop an immune response when a naive mouse is injected with so many living parasites, and during this time the parasite obviously grows and a substantial lesion forms. I thought the appearance of such massive lesions in the lung, caused by a similar immunological situation in an individual whose lung harbored *Mycobacterium tuberculosis*, might lead to pathological symptoms apparent as **type 1 tuberculosis.**

Moreover, this possibility also provided an explanation for some facts that had troubled me for years. It appears to be a rather general rule that, under most natural situations, predominant Th1 responses are either stable or the response evolves into one with a mixed Th1/Th2 phenotype, but that modulation rarely if ever occurs in the opposite direction under most conditions. It is presumably because of this direction in the evolution of the immune response that people with type 2 tuberculosis become more ill with time and do not recover.

Second, it is well known that, before the advent of antibiotics, a substantial fraction of tuberculosis patients, sent to be treated in sanatoria, spontaneously self-cured. It had always seemed to me that such spontaneous cures were not readily explicable on the hypothesis that these patients were ill as a consequence of their immune response having a debilitating Th2 component. However, mice resistant to *Leishmania major* and given a large parasite challenge form large lesions that with time resolve. The lesion grows until the Th1 immune response achieves an intensity where it kills the pathogen at a rate faster than the rate at which the pathogen is increasing by replication. A reading of the literature on tuberculosis with this framework in mind, that there were two types of tuberculosis associated with distinct types of failure by the immune system to contain the infection, no longer led to depression, but often to elation. I suggest that the establishment of mycobacterium-specific Th1 imprints will protect individuals who otherwise would develop either type 1 or type 2 tuberculosis. Individuals with a tendency to develop type 1 tuberculosis presumably would, following vaccination, make a more rapid and stronger Th1 response.

Carl Power was a graduate student at this time, working with me to establish mycobacterium-specific Th1 imprints in mice. He was stimulated by our ideas and studies on human tuberculosis to read the old tuberculosis literature. He uncovered some interesting observations and thoughts. We got copies of articles from the 1890s, written by Robert Koch, who first identified *Mycobacterium tuberculosis*, reporting on his attempts to treat patients by administering purified protein derivative (PPD) from *M tuberculosis*. Koch had imagined that he might be able to further stimulate the patient's immune system against the pathogen by giving them antigens extracted from the pathogen. We requested and received copies of these papers from the library of the US Congress. Koch's attempts at therapy greatly frustrated him.[203, 204] His therapy sometimes seemed to help the patient and sometimes greatly exacerbated disease, and hastened death, as expected of someone with type 2 tuberculosis. Carl and I realized that many of these contrary observations found a natural explanation within the framework of the Two Types of Tuberculosis Hypothesis. We thought the envisaged two types of tuberculosis were likely to be pertinent to improving tuberculosis therapy. We dreamt of writing a joint scholarly review discussing these possibilities, but never did.

I shall briefly summarize our thoughts here. People with tuberculosis are optimally treated today by administering a combination of antibiotic drugs that kills the pathogen. Treatment is required for months and is often logistically difficult to complete, thereby giving rise to drug-resistant strains of *M tuberculosis*. The treatment directly targets the bacterial pathogen and there is no attempt to harness the patient's own protective immunity. We thought therapies that attempt to optimally harness this immunity might lead to shorter and more effective treatment. Moreover, it seems most likely that the optimization of this process will be different in people ill with tuberculosis because their immune system has failed in different ways.

We have outlined above how AIDS patients may be treatable shortly after seroconversion by administering HAART to optimally modulate the anti-HIV response back from a mixed Th1/Th2 to a predominant Th1 mode. The same strategy might perhaps be effective in patients with type 2 tuberculosis. Once the Th2 component has become minor, as assessed by a small IgG_1/IgG_2 ratio, it may be possible to stop treatment and allow the patient's immune system to take care of the infection. This would mean the proposed treatment is similar in form to that now employed to treat visceral leishmaniasis. It would include monitoring of the Th1/Th2 phenotype of the response against *M tuberculosis* by the IgG isotype methodology to realize a personalized and effective treatment. Treatment of visceral leishmaniasis patients with anti-parasitic drugs is short, for a period of three weeks, and most treated individuals are resistant to reinfection.

I find it helpful to consider, on the basis of our knowledge of the immune system, why there might be different forms of tuberculosis, reflecting different types of failure by the immune system. Moreover, I suggest that such an understanding may facilitate the design of more effective treatments.

We have seen in chapter 7 that different strains of mice respond very differently to a standard challenge of *Leishmania major*. It is useful to define a "transition number" of parasites for each mouse strain. When considering the nature of immune responses following infection by a given route in mice belonging to different strains, we know that infection of mice with a number below the transition number, for the strain to which the mouse belongs, results in a stable, long term Th1 response; with a number above the transition number in a response with a substantial Th2 component. We have seen, again in chapter 7, that the transition number for *Leishmania major* can differ by a factor of as much as 10^5 in different mouse strains. Consider the consequences of infection by *Mycobacterium tuberculosis* in people for whom the transition number for such an infection varies by 10^5 or more.

Remember in addition that 90%-95% of people infected with *Mycobacterium tuberculosis* do not become ill, presumably because they generate a substantial Th1 response soon, within perhaps a month or so, after infection, and so can contain the pathogen at a low level. We have suggested that the immune response in some tuberculosis patients, with type 2 tuberculosis, cannot contain the infection due to the development of a substantial Th2 component of the response. Such individuals are likely to be those who have a low transition number for *M tuberculosis* and who are infected with a relatively large inoculum of bacteria. In this case, they would not develop a sustained and stable Th1 response.

Most infected individuals make a sufficiently strong and rapid Th1 response that the infection is contained at a low level and non-pathological level. We have argued that in some individuals, in contrast, the Th1 response is made too slowly to initially contain the infection. The infection expands and becomes substantial. The substantial Th1 response that eventually develops causes pathology by the formation, as we have seen, of granulomas in the case of lung infections. What types of people would develop such weak responses and so become type 1 tuberculosis patients? It seems highly likely that these would be individuals with relatively very high transition numbers compared to the large majority of individuals that make a substantial Th1 response sufficiently rapidly, so as to contain the infection. Suppose the transition number is 10^4 for such an individual that develops type 1 tuberculosis. We have seen that the tempo of the response is highly dependent upon the antigen load or the number of replicating organisms employed for infection, see for example Figure 8 of chapter 1. Thus, such an individual might generate an optimal Th1 response when infected with 5×10^3 mycobacteria. However, if infected with only a hundred mycobacteria, the bacteria will likely grow unimpeded until they reach a level of say 10^3 mycobacteria, when the immune system starts mounting a small response. Consider now an individual with a ten-fold lower transition number; this individual will start producing a response when there is a bacterial load of about a hundred mycobacteria. The inference is that people with type 1 tuberculosis are likely to have high transition numbers. Many likely do not have a sufficient mycobacterial burden to generate an optimal Th1 response at the time of diagnosis, unless they have reached a stage where "spontaneous" remission occurs.

To recall, patients with type 1 tuberculosis likely have a bacterial burden that is too low to optimally stimulate Th1 immunity, which is why they are ill. In this case, the standard antibiotic treatment now given is expected to reduce bacterial load and so further undermine the patient's generation of endogenous and effective immunity against the pathogen! In this case, it may well be feasible to develop a strategy of administering mycobacterial antigens in conjunction with antibiotic therapy that leads to much shorter and effective treatment. The consequences of administering such mycobacterial antigen, on the anti-microbacterial immune response, could be monitored, and the amount of antigen administered adjusted so that the patient does not develop a substantial Th2 component to their immune response. Such monitoring could be achieving by assessing the relative prevalence of IgG_1 and IgG_2 isotypes among anti-mycobacterial IgG antibodies. Such a personalized treatment seems simple to me.

Trying to explore the Ultra Low Dose BCG Vaccination Strategy

The Lung Association of Saskatchewan has a long history of involvement in human health. It arose from the Saskatchewan Anti-Tuberculosis League, which was founded in 1911. Our tuberculosis research was funded for a few years by the association, for which I was and am still especially grateful. I am only disappointed their support did not lead to a formal low dose BCG trial. The association's support allowed me to explore initiatives. For example, CEO Brian Graham used his contacts so that I could visit Honduras and meet with diverse parties, including senior health administrators and physicians, in an attempt to set up a neonatal, ultra low dose BCG vaccination trial. In the end the pieces did not all come together. I am still hopeful that this account of our tuberculosis research might lead to such a trial.

Cancer

We limit our considerations to those cases where the Th2-Skewing Hypothesis of tumor escape applies. We have argued above that this hypothesis is true of two murine tumor systems we have examined, and likely to be true in many other murine tumor systems. There is also indirect evidence that it may apply in human cancer.[205-209]

Vaccination against cancer

The implications for strategies of vaccination are straightforward within our framework, and have already been illustrated in considering how to achieve effective vaccination against HIV-1 and *M. tuberculosis*. The low dose vaccination strategy would appear to provide a way of immunizing against cancer, particularly in view of other developments in the field of cancer immunology. The investigations by Thierry Boon and his colleagues have defined many tumor-associated antigens recognized by CTL[8] and others have defined similar antigens recognized by antibodies.[7] Such CTL and antibodies are more prevalent in cancer patients than in their normal counterparts, giving rise to the belief that the immune system provides surveillance against cancer. Tumors of the same type, and even of different types, often share tumor-associated antigens. This finding is important both in trying to prevent cancer by vaccination and in its immunotherapeutic treatment. Again, it would seem most worthwhile to examine whether vaccination with low numbers of BCG vectors, encoding the most frequently expressed tumor-associated antigens, may be effective in preventing cancer.

Immunotherapy of cancer

The two most general ways of treating cancer are to partially remove the growth, thereby lowering the antigen load, and giving agents or treatments that kill dividing cells. It is remarkable that both forms of treatment have the potential for modulating mixed Th1/Th2 responses to a Th1 mode. We have seen, in the case of human visceral leishmaniasis,[176] that lowering the antigen load modulates a mixed Th1/Th2 response to a Th1 mode.

Radiation, as well as giving drugs that kill dividing cells such as cylcophosphamide, are used in cancer treatment. Both these manoeuvres have the potential for acting not only on the tumor directly, but also on the cells of the immune system, thereby potentially modulating immune responses from a humoral to a cell-mediated mode. North showed that whole body irradiation of a mouse with a progressively growing and established meth A fibrosarcoma could result in tumor regression. He also showed that radiation was effective due to its action on tumor-specific CD4 T cells and not because it directly affected tumor growth.[183] North and Awwad also showed that the administration of cyclophosphamide to an animal, a day before an inoculation of tumor cells that normally resulted in progressive tumor growth, could result in tumor regression. It is known that the administration of cyclophosphamide a day before tumor implantation means it does not act to directly inhibit the growth of the tumor cells themselves. Cyclophosphamide must act to deplete the recipient of some cell involved in suppressing a protective response. North reconstituted the cyclophosphamide-treated recipient one day after cyclophosphamide administration with spleen cells from normal mice to see what cell type was involved in suppressing the protective response. The incriminating cell was found to be a CD4 T cell.[210] It so happens that others had shown that the administration of the same dose of cyclophosphamide, as used by North and given in the same manner, totally inhibits an antibody response to SRBC and that the mice produce a potent DTH response against SRBC instead.[76] All these observations are readily understood in terms of the Threshold Hypothesis, as the generation of Th2 responses requires stronger CD4 T cell cooperation than does the generation of Th1 responses.

It seems to me that, if there is a simple way to longitudinally monitor the Th1/Th2 phenotype of the anti-cancer immune response, treatment could then be adjusted to optimally harness the patient's protective response against the tumor. Current standard treatments are not directed at harnessing the patient's own protective immunity. We have provided grounds for thinking that the relative preponderance of different IgG isotypes present among anti-tumor antibodies can be used to longitudinally monitor the Th1/Th2 phenotype of the anti-tumor immune response, and thus the value of this parameter could be used to guide treatment.[10, 182, 185] We are currently exploring this possibility in human colorectal cancer.

Concluding Remarks

To harvest the insights of inspiration requires a forward looking and optimistic cast of mind. The proposals outlined above could only be envisaged in such a state. Their exploration will require stoicism, characteristic of this phase of scientific enquiry.

I would like to finish my scientific story by commenting on my perception of the importance of, and the difficulty of sustaining, an inspirational outlook in contemporary times. Experiences during the first half of my immunological career made me aware of how easy it is to be overwhelmed by information, by sensory overload and the concomitant loss of vision. I have at times despaired during the first half of my career, feeling that understanding is impossible. I lost faith in the value of the hypotheses I had been developing and of thinking in an elemental way. I had at times to cut myself off from exposure to current fashions and their propaganda, to allow the regeneration of my belief that nature is comprehensible, and to escape the despair I experienced from the disintegration of my inspiration. I could then feel again the value of seeking the big picture. I believe in the big picture—believing in it is for me the means of being most significantly constructive. Surprisingly, as the information load became more severe over the decades, it became ever more evident to me that the only way to deal with this load was to pay ever more respect to subtle argument, to pay attention to my gut feelings, and always to strive to address the most central questions. I believe that the story I have told bears on contemporary existential questions we all face as we seek ways of discovering meaning in what we do.

Glossary

Technical terms are listed alphabetically, followed in parentheses by the
page where the term is ***highlighted*** and its meaning indicated by context

activatable (36)

activating receptors (197)

active immunization (9)

acute inflammation (6)

adaptability (7)

adjuvants (38, 236)

acquired immune deficiency
     syndrome (AIDS) (3)

alleles (100)

allergens (19)

allergic (2,21)

allergy (19)

anaphylaxis (193)

antibodies (10)

antibody effector cells (58)

antibody precursor cell (57)

antigen (11)

antigen, foreign (2)

antigen, self (2)

Antigen-Binding Model, Simple (53)

antigen-binding site (19)

Antigen Bridge Model of the B cell/T
     helper cell interaction (64)

antigenic (11)

antigen presentation (107, 149)

antigen presenting cells (107, 149)

antigen processing (106, 149)

antigen-pulsed macrophages (105)

antigen-sensitive cell (39)

antigens, polymeric (63)

anti-inflammatory immunity (141, 143)

APC (107)

asthma (19)

attenuated (8)

autoantibodies (69)

autocrine (113)

autoimmune diabetes (3)

autoimmune hemolytic anemia (2)

autoimmunity (3, 96)

autoreactivity (3, 96)

bacille Camille Gueron (BCG) (213)

bacterial lysis (10)

BALB/c mice (205)

B cells (63)

B cell hybridomas (114)

BCG (213)

binding site, antigen (19)

bone marrow cells (63)

bovine serum albumin (48, 53)

BSA (48, 53)

CD40 ligand (123)

CD4, CD8 (72)

CD40 (123)

CD40L (123)

cell-mediated immune deviation (77, 80)

cell-mediated immunity (3, 22, 75)

cell-mediated imprints (4)

central tolerance (117)

CFA (129, 167)

CGG (62)

CH_1, CH_2, CH_3, C_L (19)

chemokines (165)

chicken γ-globulin (62)

classes of immunity (21)

clones, T cell (114)

coherence (231)

coherent (181)

References and Notes

1. P Ehrlich and J Morgenroth, 1901. Uber Hamolysine: Funfte Mittheilung. Berl Klin. Wschr. English tranlation in The Collected Papers of Paul Ehrlich, Vol. 1 1956. London and New York: Pergamon Press pp246-255

2. W Damashek and RS Schwartz, 1938. The presence of hemolysins in acute hemolytic disease. *N Eng J Med* 218:75-80

3. M Jutel, M Akdis, F Budak, C Aebischer-Casaulta, M. Wrzyszcz, K Blaser and CA Akdis, 2003. IL-10 and TGF-β cooperate in the regulatory T cell response to mucosal allergens in normal immunity and specific immunotherapy. *Eur J Immunol.* 33:1205

4. F Pociot et al, 2010. Genetics of type 1 diabetes: What's next? *Diabetes* 59: 1561

5. KR Fowke, NJD Nagelkerke, J Kimani, JN Simonsen, AO Anzala, et. al., 1996. Resistance to HIV-1 infection among prostitutes. *Lancet* 348: 1347

6. PA Bretscher, N Ismail, JN Menon, CA Power, J Uzonna and G Wei, 2001. Vaccination against and treatment of tuberculosis, the leishmaniases and AIDS: perspective from basic immunology and immunity to chronic intracellular infections. *Cell Mol Life Sci* 58:1879

7. G Li, A Miles, A Line, and RC. Rees, 2004. Identification of tumor antigens by serological analysis of cDNA expression cloning. *Cancer Immunol. Immunother.* 53:139

8. T Boon, JC Cerottini, B Van den Eynde, P van der Brugen A Van Pel, 1994. Tumor antigens recognized by T lymphocytes. *Ann Rev Immunol* 12: 33

9. P Ehrlich, 1909. Ueber den jetzigen Stand der Karzinomforschung. *Ned. Tijdschr Geneeskd* 5:273.

10. D Hamilton and PA Bretscher, 2008. The commonality in the regulation of the immune response to most tumors: the prevalence of immune class

deviation as a tumor escape mechanism and its significance for vaccination and immunotherapy *Canc Ther* 6: 745

11. MC Dinauer and SH Orkin, 1988. Chronic granulomatous disease. Molecular genetics. *Hematol Oncol Clin North Am* 2:225

12. JH Humphrey and RG White, 1970. Immunology for Medical Students, Blackwell Scientific Publications, Oxford and Edinburgh, Third Edition
    Much of the account given here of the history of immunology, up to the early 1900s, is modeled on this classic text.

13. RC Valentine, 1967. Electron microscopy of an antibody-hapten complex. *J Mol Biol* 27: 615

14. K Landsteiner, 1945. The Specificity of Serological Reactions. Harvard University Press, Cambridge, Mass

15. A Aderem and RJ Ulevitch, 2000. Toll-like receptors in the induction of the innate immune response. *Nature* 406: 782

16. RR Porter, 1991. Lecture for the Nobel Prize for physiology or medicine 1972: Structural studies of immunoglobulins. *Scand J Immunol* 34: 381

17. K Landsteiner and MW Chase, 1942. Experiments on transfer of cutanous sensitivity to simple compounds. *Proc Soc Exp Biol Med* 49:688.

18. J Cerottini and KT Brunner, 1974. Cell-Mediated Cytotoxicity, Allograft Rejection, and Tumor Immunity. *Adv Immunol* 18: 67

19. Salvin SB, 1958. Occurrence of DTH during the development of Arthus type sensitivity. *J Exp Med* 107:109

20. T Godal, 1974. The role of immune response to mycobacterium leprae in host defense and tissue damage in leprosy. *Prog Immunol* 11: 4

21. PEM Fine, 1988. BCG Vaccination against Tuberculosis and Leprosy. *Br. Med. Bull.* 44: 29

22. C Havele, 1980. Immunological Tolerance and Autoimmunity, PhD Thesis, University of Alberta

23. NC Peters, 2003. A Requirement for CD4 T-helper cell cooperation in the generation of Immune Responses: Implications for Models of pTh cell Activation and Self-Nonself Discrimination. PhD Thesis, University of Saskatchewan

24. P Ehrlich, 1900. On immunity with special reference to cell life (Croonian Lecture) *Proc Roy Soc London,* 66: 424

25. L Pauling, 1940. A Theory of the Structure and Process of Formation of Antibodies. *J Am Chem Soc* 62: 2643

26. FM Burnet and F Fenner, 1949. The Production of Antibodies, McMillan and Co., New York

27. RE Billingham, L Brent and PB Medawar, 1953. Actively acquired tolerance of cells. *Nature* 172: 603

28. NK Jerne, 1955. The Natural Selection Theory of Antibody Formation, *Proc Natl Acad Sci* 41: 849

29. FM Burnet, 1959. The Clonal Selection Theory of Acquired Immunity, Cambridge University Press

30. DW Talmage, 1957. Allergy and Immunology. *Ann Rev Med* 8:239

31. JD Watson and FHC Crick, 1953. A Structure for Deoxyribose Nucleic Acid *Nature* 171: 737

32. F Sanger, 1964. The chemistry of insulin, in Nobel Lectures, Chemistry 1942-1962, Elsevier Pub. Co., Amsterdam.

33. CB Anfinsen, E Haber,* M Sela, and FH White, 1961. The kinetics of formation of native ribonucleasse during oxidation of the reduced polypetide chains. *Proc Natl Acad Sci* 47: 1309

34. J Lederberg, 1959. Genes and antibodies. *Science* 129: 1649

35. S Brenner and C Milstein, 1966. Origin of antibody variation. *Nature* 211: 242

36. I Green, WE Paul and B Benacerraf, 1966. The behavior of hapten-poly-L-lysine conjugates as complete antigens in genetic responder and as haptens in genetic nonresponder guinea pigs. *J Exp Med* 123:859

37. B Benacerraf, I Green and WE Paul, 1967. The Immune Response of Guinea Pigs to Hapten-Poly-L-Lysine Conjugates as an Example of the Genetic Control of the Recognition of Immunogenicity. *Cold Spring Harbor Symp Quant Biol.* 32: 569.

38. Immunological Tolerance, Proceedings of an International Conference, at Brook Lodge, September 18-20, 1968, Editors M Landy and W Braun, Academic Press, New York, London, 1969

39. PA Bretscher and M Cohn, 1968. Minimal Model for the Mechanism of Antibody Induction and Paralysis by Antigen. *Nature* 220: 444

40. DW Dresser and NA Mitchison, 1968. The mechanism of immunological paralysis. *Adv Immunol* 8: 129

41. *Cold Spring Harbor Symp Quant. Biol.*, 1967, 32

42. NA Mitchison 1967. Anigen Recognition Responsible for the Induction In Vitro of the Secondary Response. *Cold Spring Harbor Symp Quant Biol* 32: 431.

43. K Rajewsky and E Rottlander 1967. Tolerance Specificity and the Immune Response to Lactic Dehydrogenase Isoenzymes. *Cold Spring Harbor Symp Quant Biol* 32: 547.

44. M Hasek, A Lengerova and T Hraba, 1961. Transplantation immunity and tolerance. *Adv Immunol* 1:1,

45. WO Weigle, 1961. The immune response of Rabbits Tolerant to Bovine Serum Albumin to the Injecion of Other Heterologous Serum Albumins. *J Exp Med* 114: 111

46. FC Frei, B Benacerraf, GJ Thornbecke 1965. Phagocytosis of the antigen, a crucial step in the induction of a primary response. *Proc Natl Acad Sci* 53: 20

47. PA Bretscher and M Cohn, 1970. A theory of Self-Nonself Discrimination. *Science* 169: 1042

48. PA Bretscher, 1972. The Control of Humoral and Associative Antibody Synthesis. *Transplantation Reviews* 11: 217

49. HN Claman, EA Chaperon and RF Triplett, 1966. Thymus, marrow cell combinations-synergism in antibody production. *Proc Soc Exp Biol Med* 122: 1167

50. GF Mitchell and JFAP Miller, 1968. Cell to cell interaction in the immune response. II The source of the hemolysin-forming cells in irradiated mice given bone marrow cells and thymus or thoracic duct lymphocytes. *J Exp Med* 128: 821

51. NA Mitchison, 1971. The carrier effect in the secondary response to hapten-protein conjugates. II. Cellular cooperation. *Eur J lmmunol* 1:18.

52. MC Raff, 1970. Role of thymus-derived lymphocytes in the secondary humoral immune response in mice. *Nature* 226: 1257.

53. M Cohn, 1994. The Wisdom of Hindsight. *Ann Rev Immunol* 12:1

54. WO Weigle, 1973. Immunological Unresponsiveness. *Adv Immunol* 16:61

55. JB Zabriski, KC Hsu, and BC Seegal, 1982. Heart-reactive antibody associated with rheumatic fever: characterization and diagnostic significance. *Clin Exp Immunol* 7: 145

56. PJ McCullagh, 1972. The abrogation of immunological tolerance by means of allogeneic confrontation. *Trans Rev* 12: 180

57. RK Gershon and K Kondo 1971. Infectious Immunological Tolerance. *Immunol* 21: 903

58. J Salk, PA Bretscher, PL Salk, M Clerici and GM Shearer, 1993. A Strategy for Prophylactic Vaccination against AIDS. *Science*, 260: 1269

59. PA Bretscher, 1974. Hypothesis: On the control between cell-mediated, IgM and IgG immunity. *Cell Immunol* 13: 171

60. PA Bretscher, 2014. On the Mechanism Determining the Th1/Th2 Phenotype of the Immune Response, and its Pertinence to Strategies for the Prevention, and Treatment of, Certain Infectious Diseases. *Scand J Immunol* 79:361

61. GL Asherson and SH Stone, 1965. Selective and specific inhibition of 24 hr skin reactions in the guinea pig. I. Immune Deviation: description of the phenomenon and the effect of splenectomy. *Immunol* 9: 205

62. CR Parish, 1972. The relationship between humoral and cell-mediated immunity. *Trans Rev* 13: 35

63. P Bretscher, 2000. Contemporary models for peripheral tolerance and the classical 'historical postulate'. *Semin Immunol* 12: 221; discussion 12: 257

64. NA Mitchison, 1967. Immunological paralysis as a dosage phenomenon, In *Regulation of the Antibody Response* (Ed. B Cinader) pp54-67. CC Thomas, Springfield, Ill.

65. MN Pearson and S Raffel, 1971. Macrophage-digested antigen as inducer of delayed hypersensitivity. *J Exp Med* 133: 494

66. JH Humphrey & RR Dourmashkin, 1969. The lesions in cell membranes caused by complement. *Adv Immunol* 11: 75

67. I Hellstrom and KE Hellstrom, 1969. Studies on cellular immunity and its serum-mediated inhibition in Moloney-virus-induced mouse sarcomas. *Int J Cancer*, 4: 587

68. G Klein, 1968. Tumor-specific transplantation antigens: G.H.A. Clowes Memorial Lecture. *Can Res* 28: 625

69. PA Bretscher, 1981. Significance and mechanisms of cellular regulation of the immune response. *Fed Proc* 40:1473

70. RK Gershon and K Kondo, 1972. Tolerance to sheep red blood cells: breakage with thymocytes and horse red cells. *Science* 175: 996

71. PH Lagrange, GB Mackaness and TE Miller, 1974. Influence of dose and route of antigen injection on the immunological induction of T cells. *J Exp Med* 139: 528

72. J Lesley, R Hyman and G Dennert 1974. The effect of antigen density on complement-mediated lysis, T cell-mediated killing and antigenic modulation, *J Nat Can Inst* 53: 1759

73. PA Bretscher, 1977. An Integration of B and T cells in Immune Activation, in B and T cells in Immune Recognition, F Loor and G Roelants, Eds, John Wiley and Son, pp457-485

74. P Debre, JA Kapp, MF Dorf and B Benacerraf, 1975. Genetic control of specific immune suppression. I Experimental conditions for the stimulation of suppressor cells by the copolymer GT in non-responder BALB/c mice. *J Exp Med* 142: 1447

75. JFAP Miller and GF Mitchell 1970. Cell to cell interaction in the immune response. V Target cells for tolerance induction *J Exp Med* 131:675

76. PH Lagrange, GB Mackaness and TE Miller, 1974. Potentiation of T cell-mediated immunity by selective suppression of antibody formation with cyclophosphamide. *J Exp Med* 139:1529.

77. IA Ramshaw, PA Bretscher and CR Parish, 1977. Regulation of the immune response. II Repressor T cells in cyclophosphamide-induced tolerant mice. *Eur J Immunol* 7: 180

78. Ramshaw IA, Bretscher PA, Mckenzie IFC, Parish CR. 1977. Discrimination of suppressor T cells of humoral and cell-mediated immunity by anti-Ly and anti-Ia sera. *Cell Immunol.* 31: 364

79. IA Ramshaw, PA Bretscher and CR Parish. 1976. Regulation of the immune response. I Suppression of delayed-type hypersensitivity by T cells from mice expressing humoral immunity. *Eur J Immunol* 6: 674

80. G Wiedermann, H Denk, H Stemberger, R Eckersforfer and G Tappenheimer, 1975. Influence of the antigenicity of target cells on the antibody-mediated cytotoxicity of nonsensitized lymphocytes *Cell Immunol* 17: 440

81. RM Zinkernagel and PC Doherty, 1975. H-2 compatability requirement for T-cell-mediated lysis of target cells infected with lymphocytic choriomeningitis virus. Different cytotoxic T-cell specificities are associated with structures coded for in H-2K or H-2D *J Exp Med* 141: 1427

82. R Shimonevitz, J Kappler, P Marrack, and H Grey, 1983. Antigen recognition by H-2 restricted T cells. I Cell-free processing. *J Exp Med* 158:303

83. S Buus, A Sette, SM Colon and HM Grey, 1987. The relationship between MHC restriction and the capacity of Ia to bind immunogenic peptides. *Science* 235:1353

84. A Lanzavecchia, 1985. Antigen-specific interaction between B and T cells. *Nature* 314: 537

85. S Tonegawa, 1983. Somatic generation of diversity. *Nature* 302: 575

86. JJ Moon, HH Chu, M Pepper, SJ McSorley, SC Jameson, RM Kedl, and MK Jenkins, 2007. Naive CD4(+) T cell frequency varies for different epitopes and predicts repertoire diversity and response magnitude. *Immunity* 27: 203

87. DA Nemazee and K Burki, 1989. Clonal Deletion of B lymphocytes in a transgenic mouse bearing anti-MHC class I genes. *Nature* 337: 562

88. CC Goodnow, J Crosbie, H Jorgensen, RA Brink and A Basten, 1989. Induction of self-tolerance in mature peripheral B lymphocytes. *Nature* 342:385

89. G Kohler and C Milstein, 1975. Continuous cultures of fused cells secreting antibody of predefined specificity. *Nature* 256: 495

90. JW Kappler, N Roehm and P Kappler, 1989. T cell tolerance by clonal elimination in the thymus. *Cell* 49: 273

91. RM Zinkernagel, GN Callahan, A Althage, S Cooper, PA Klein and J Klein, 1978. On the thymus in the differentiation of "H-2 self-recognition" by T cells: evidence for dual recognition? *J Exp Med* 147: 882

92. A Bhushan, and LR Covey, 2002. CD40:CD40L interactions in X-linked and non-X-linked hyper-IgM syndromes. *Immunol Res* 24: 311

93. Metcalf, E. S. & Klinman, N. R. In vitro tolerance induction of neonatal murine B cells. *J Exp Med* 143, 1327-40 (1976).

94. CC Goodnow, 1992. Transgenic mice and analysis of B-cell tolerance. *Ann Rev Immunol.* 10: 489

95. JA Keene and J Forman. 1982. Helper activity is required for the in vivo generation of cytotoxic T cells. *J Exp Med* 155: 768

96. S Guerder and P Matzinger, 1992. A fail-safe mechanism for maintaining self tolerance. *J Exp Med* 176: 553

97. MJ Tucker and PA Bretscher, 1982. T cells cooperating in the induction of delayed-type hypersensitivity act via the linked recognition of antigenic determinants. *J Exp Med* 155: 1037

98. PA Bretscher, 1986. A Cascade of T-T Interactions, Mediated by the Linked Recognition of Antigen, in the Induction of T cells able to Help Delayed-Type-Hypersensiiviuty Responses. *J Immunol* 137: 3726

99. M Gerloni, S Xiong, S Mukerjee, SP Schoenberger, M Croft and M Zanetti, 2000. Functional cooperation between T helper determinants. *Proc Natl Acad Sci.* 97: 13269

100. KJ Lafferty, SJ Prowse and J Charmaine, 1983. Immunobiology of Tissue Transplantation: A Return to the Passenger Leucocyte Concept. *Ann Rev Immunol* 1:143

101. H Quill and RH Schwartz, 1987. Stimulation of normal inducer T cell clones with antigen presented by purified Ia molecules in planar lipid membranes: specific induction of a long-lived state of proliferative nonresponsiveness. *Immunol* 138: 3704

102. CA Janeway, Jr., 1989. Approaching the asymptote? Evolution and revolution in immunology. *Cold Spring Barb Symp Quant Biol* 54 Pt 1, 1

103. P Matzinger, 1994. Tolerance, danger, and the extended family. *Annu Rev Immunol* 12: 991

104. DL Kaufman, M Clare-Salzler, J Tian, T Forsthuber, GSP Ting, P Robinson, MA Atkinson, EE Sercarz, AJ Tobin, and PV Lehmann. 1993. Spontaneous loss of T cell tolerance to glutamic acid decarboylase in murine insulin-dependent diabetes. *Nature* 366: 69

105. VK Tuohy, M Yu, L Yin, JA Kawczak, RP Kinkel, 1999. Epitope-spreading: spontaneous regressions of primary autoreactivity during chronic progression of EAE and MS. *J Exp. Med.* 189:1033-1042

106. P Waterhouse, JM Penninger, E Timms, et al, 1995. Lymphoproliferative disorders with early death in mice deficient in CTLA-4. *Science* 270: 985

107. BT Fife, M Gupels Bupp, TN Eagar, et al, 2006. Insulin-induced remission in new-onset NOD mice is maintained by the PD-1-PD-1L pathway. *J Exp Med* 203:883

108. T R Mosmann and R L Coffman, 1989. TH1 and TH2 Cells: Different Patterns of Lymphokine Secretion Lead to Different Functional Properties *Ann Rev Immunol* 7: 145

109. CS Hsieh, SE Macatonia, CS Tripp, SF Wolf, A O'Garra, A and KM Murphy, 1993. Development of TH1 CD4 T cells though IL-12 produced by Listeria-induced macrophages. *Science* 1993; 260:547-49

110. SL Swain, AD Weinberg, M English and G Huston, 1990. IL-4 directs the development of Th2-like helper effectors. *J Immunol* 145:3796

111. TF Gajewsky and FW Fitch, 1988. Anti-prolierative effect of IFN-γ in immune regulation. I IFN-γ inhibits the proliferation of Th2 but not Th1 HTL clones. *J Immunol* 140:4245

112. SL Reiner and RM Locksley, 1995.The regulation of immunity to *Leishmania major. Ann Rev Immunol* 13:151

113. MD Sadick, FP Heinzel, BJ Holaday, RT Pu, RS Dawkins and RM Lockseley, 1999. Cure of murine leishmanisis with anti-IL4 monoclonal antibody. *J Exp Med* 1999; 171:115-127

114. DT Fearon and RM Locksley, 1996. The instructive role if innate immunity in the acquired immune response. *Science*, 1996; 272:50-54

115. A Iwasaki, R Medzhitov, 2015. Control of adaptive immunity by the innate immune system. *Nature Immunol* 16: 343

116. M Moser and KM Murphy, 2000. Dendritic cell regulation of TH1-TH2 development. *Nat Immnol* 1: 199

117. PA Bretscher, 1987. Requirement for antigen in lipopolysaccharide-dependent induction of B cells. *Eur J Immunol* 8: 534

118. SJ Challacombe and TB Tomasi, 1980. Systemic tolerance and secretory immunity after oral immunization. *J Exp Med* 152: 1459

119. M van den Nent Kolfscholten, J Schuurman, M Losen, WK Bleeker, P Martinez, A Vermeulen, TH den Bleker, L Wiegman, T Vink, LA Aarden, MC be Baets, JgJ van de vWinkel, RC Aalberse and PWHI Parren. 2007. Anti-inflammatory activity of human IgG$_4$ antibody by dynamic Fab arm exchange. *Science* 317:1554-1557

120. J Meiler, J Zumkehr, S Klunker, B Ruckert, CA Akdis and M Akdis, 2008. In vivo switch to IL-10-producing T regulatory cells in high dose allergen exposure. *J Exp Med* 205: 2887

121. S Sakaguichi, T Takahashi and Y Nishizuka, 1982. Study on cellular events in post-thymectomy autoimmune oophoritis in mice. I Requirement for Lyt-1 effector cells for oocytes damage after adoptive transfer. *J Exp Med* 156: 1565

122. K Aschenbrenner, LM D'Cruz, EH Vollman, M Hinterberger, J Emmerlich, KL Swee, A Rolink and L Klein, 2007. Selection of Foxp3+ regulatory T cells specific for self antigen expressed and presented by Aire+ medullary thymic epithelial cells. *Nature Immunol* 8: 351

123. CM Sun, JA Hall, RB Blank, N Bouladoux, M Oukka, JR Mora and Y Belkaid, 2007. Small intenstine lamina propria dendritic cells promote de novo generation of Foxp3 T reg cells via retinoic acid. *J Exp Med* 204: 1775

124. M Akdis, J Verhagen, A Taylor, F Karamloo, C Karagiannis, R Crameri, S Thunberg, G Deniz, R Valenta, H Fiebig, C Kegel, R Disch, CB Schmidt-Webert, K Blaser, and CA   Akdis, 2004. Immune responses in healthy and allergic individuals are characterized by a fine balance between allergic-specific T regulatory, T helper 1 and T helper 2 cells. *J Exp Med* 199:1567

125. NK Jerne, 1974. Towards a Network Theory of the Immune System. *Ann. Immunol. (Inst. Pasteur)* **125C**: 373

126. G Petrányi, 1981. Nobel Prize winners in medicine for 1980. Immunogenetic significance of the main histocompatibility system. <u>*Orvosi hetilap*</u> 122: 835

127. ME Dorf and B Benacerraf, 1984. Suppressor cells and immunoregulation. *Ann Rev Immunol* 2: 127

128. NK Jerne, 1984. <u>Nobel lecture: The Generative Grammar of the Immune System</u>

129. N Ismail and PA Bretscher, 1999. The Th1/Th2 nature of concurrent immune responses to unrelated antigens can be independent. *J Immunol* 163: 4842

130. PA Bretscher, 1999. A Two Step, Two Signal Model for the primary activation of precursor helper T cells. *Proc Natl Acad Sci* 96: 185

131. OT Chan, LG Hannum, AM Haberman, MP Madaio, and MJ Shlomchik, 1999. A novel mouse with B cells but lacking serum antibody reveals an antibody-independent role for B cells in murine lupus. *J Exp Med* 189: 1639

132. DL Lenshow, TL Walunas and JA Bluestone, 1996. The CD28/B7 system of T cell costimulation. *Ann Rev Immunol* 14: 233

133. D Rossi and A Zlotnik, 2000. The Biology of Chemokines and their Receptors. *Ann Rev Immunol* 18: 217

134. CA Power, CL Grand, N Ismail, NC Peters, DP Yurkowski and PA Bretscher, 1999. A valid ELISPOT assay for enumeration of ex-vivo, antigen-specific IFN-gamma-producing T cells. *J Immunol Methods* 227: 99

135. N Peters, D Hamilton and PA Bretscher, 2005. Single cell analysis of hen egg lysozyme (HEL) immunized BALB/c and CBA mice defines the peptide-specificity of all the HEL-specific Th cells generated and reveals different repertoires of the cells obtained from different lymphoid organs and from one organ but producing different cytokines. *Eur J Immunol* 33: 55

136. NC Peters, DR Kroeger, S Mickelwright and PA Bretscher, 2009. CD4 T cell cooperation is required for the in vivo activation of CD4 T cells, Int Immunol 11: 1213

137. Hawiger, D., et al., 2001. Dendritic cells induce peripheral T cell unresponsiveness under steady state conditions in vivo. *J Exp Med,* 194: 769

138. RM Steinmann, 2003. Dendritic cell function in vivo during the steady state: A role in peripheral tolerance *Ann N Y Acad Sci* 987: 15-25

139. PA Bretscher, 2014. The Activation and Inactivation of Mature CD4 T cells: A Case for Peripheral Self-Nonself Discrimination. *Scan J Immunol* 79:48

140. JC Edwards, L Szczepanski, J Szczepanski, A Filipowicz-Sosnowska, T Emery, DR Close, RM Stevens and TShaw, 2004. Efficacy of B cell targeted therapy with rituximab in patients with rheumatoid arthritis. *N Eng J Med* 350: 2572

141. DR Kroeger, CR Rudulier and PA Bretscher. 2013. Antigen presenting B cells facilitate CD4 T cell cooperation resulting in enhanced effector and memory CD4 T cells. *PLoS One,* 8:e7734654

142. M Cohn, 2015. Thoughts engendered by Bretscher's Two-Step, Two-Signal Model for a Peripheral Self-Nonself Discrimination and the Origin of Primer Effector T Helpers. *Scand J Immunol* 81:87

143. PA Bretscher, 2015. A conversation with Cohn on the Activation of CD4 T cells. *Scand J Immunol* 82:147

144. RI Mishell and RW Dutton, 1967. Immunization of disassociated spleen cell cultures of normal mice. *J Exp Med* 126: 42

145. IA Ramshaw and D Eidinger, 1979. The in vitro induction of T cells which mediate delayed type hypersensitivity toward horse red blood cells. *Cell Immunol* 42: 4247

146. PA Bretscher, 1983. In vitro analysis of the cellular interactions between unprimed lymphocytes responsible for determining the class of response an antigen induces: specific T cells switch a cell-mediated response to a humoral response. *J Immunol* 131: 1103

147. PA Bretscher, 1983. Regulation of the immune response induced by antigen. I Specifc T cells switch the in vivo response from a cell-mediated to humoral mode. *Cell Immunol* 81: 345

148. N Ismail and P Bretscher, 2001. More antigen-dependent CD4+ T cell/CD4+ T cell Interactions are required for the Primary Generation of Th2 than of Th1 cells, *Eur J Immunol* 31: 1765

149. Christopher D. Rudulier, David R Kroeger and Peter A. Bretscher, 2014. Stronger CD4 T cell interactions, mediated by B7/CD28 interactions, are required to generate Th2 than Th1 responses. *J Immunol.* 192: 5140-5150

150. N Ismail, A Basten, H, Briscoe and PA Bretscher, 2004. Increasing the foreign-ness of an antigen, by coupling a second and foreign antigen to it, increases the Th2 component of the immune response to the first antigen. *Immunol.* 115: 34

151. JL Flynn, J Chan, KJ Triebopld, DK Dalton, TA Stewart and BR Bloom, 1993. An essential role for interferon gamma in resistance to Mycobacterium tuberculosis infection. *J Exp Med* 178: 2249

152. EM Coccia, E Stellacci, G Marziali, G Weiss and A Battistni, 2000. IFN gamma and IL-4 differentially regulate through IRF-1 modulation NO synthetase gene expression. *Int Immunol* 12:977

153. JG Perrigare, SS Saenz, ML Siracusa, EJ Allenspach, et al, 2009. MHC class II-dependent basophil-CD4 T cell interactions promote Th2 cytokine-dependent immunity. *Nature Immunol* 10: 697

154. DI Godfrey and M Kronenberg, 2004. Going both ways: immune regulation via CD1d-dependent NK T cells. *J Clin Invest* 114: 1379

155. L Liu, BE Rich, J Inobe and HL Weiner, 1998. Induction of Th2 cell differentiation in the primary immune response: dendritic cells isolated from adherent cell culture treated with IL-10 prime naïve CD4 T cells to secrete IL-4. *Int Immunol* 10: 1017

156. DF Fiorentino, MW Bond and TR Mosmann, 1989. Two types of mouse T helper cell. IV Th2 clones secrete a factor that inhibits cytokine production by Th1 clones. *J Exp Med* 170: 2081

157. I Ivanov, K Atarashi, N Manel et al, 2010. Induction of Th17 cells by segmented, filamentous bacteria. *Cell* 139:485

158. DP Strachan, 1989 Hay fever, hygiene, and household size. *Brit Med J* 299: 1259

159. AJ Bancroft, KJ Else and RK Grencis, 1994. Low level infection with *Tricuris muris* significantly affects the polarisation of the CD4 response. *Eur J Immunol* 24: 3113

160. TD Langford, MP Housely, M Boes, J Chen MF Kagnoff, FD Gillin, and FD Eckmann, 2002. Central importance of immunoglobulin A in host defense against Giardia ssp. *J Immunol* 70: 11

161. MR Karlsson, J Rugtveit and T Brandtzaeg, 2004. Allergen-responsive CD4+CD25 regulatory T cells in children who have out-grown cow's milk allergy. *J Exp Med* 199: 1679

162. W Haas, P Pereira and S Tonegawa, 1993. Gamma/Delta cells. *Ann Rev Immunol* 11: 637

163. CA Biron, KB Nguyen, GC Pien, LP Cousens, TP Salazr-Mather, 1999. Natural Killer Cells in Antiviral Defense: Function and Regulation by innate cytokines. *Ann Rev Immunol* 17:189

164. H Ljunggren and K Karre, 1990. In search of the 'missing self': MHC molecules and NK cell recognition. *Immunol Today* 11: 237

165. GF Mitchell, JM Curtis, RG Scollay and E. Handman, 1980. Resistance and abrogation of resistance to cutaneous leishmaniasis in reconstituted BALB/c nude mice. *Aust J Exp Med* Sci 58: 521

166. PA Bretscher, G Wei, JN Menon, H Bielefeldt-Ohmann, 1992. Establishment of stable cell-mediated immunity that makes "susceptible" mice resistant to *Leishmania major*. *Science* 257: 539

167. J Menon and PA Bretscher, 1998. Parasite dose determines the Th1/Th2 nature of the reponse to *Leishmania major* independently of infection route, strain of host or of parasite *Eur J Immunol*. 28: 4020

168. Dam, HG ten, 1984, Research on BCG vaccination *Adv. Tuberc. Res* 21:79

169. CA Power, G Wei and PA Bretscher, 1988. Mycobacterial dose defines the Th1/Th2 nature of the immune response independently of whether immunization is by the intravenous, subcutaneous or intradermal route. *Inf Immun* 66:5743

170. TG Kiros, CA Power, G Wei, and PA Bretscher, 2010. Immunization of newborn and adult mice with low numbers of BCG leads to Th1 responses, Th1 imprints and enhanced protection upon BCG challenge. *Immunother* 2: 25

171. PA Bretscher, JN Menon, C Power, J Uzonna and G Wei, 2001. A case for neonatal, low dose BCG vaccination. *Scand J Inf Dis* 33: 253

172. F Buddle, GW de Lisle, A Pfeffer and FE Aldwell, 1995. Immunological responses and protection against *Mycobacterium bovis* in calves vaccinated with a low dose of BCG. *Vaccine* 13: 1123

173. RG Titus, R Ceredig, JC Cerottini and JA Louis, 1985. Therapeutic effect of anti-L3T4 monoclonal antibody GK1.5 on cutaneous leishmanisis in genetically susceptible BALB/c mice. *J Immunol* 135: 2108

174. MD Sadick, FP Heinzel, BJ Holaday, RT Pu, RS Dawkins and RM Lockseley, 1990. Cure of murine leishmanisis with anti-IL4 monoclonal antibody. *J Exp Med* 171: 115

175. JE Uzonna, and PA Bretscher, 2001. Anti-IL4 antibody therapy causes regression of chronic lesions caused by medium-dose leishmania major infection, *Eur J Immunol* 31: 3175

176. A Hailu, JN Menon, N Berhe, L Gedamu, TH Hassard, PA Bretscher, 2001. Distinct immunity in visceral leishmaniasis patients from that in subclinically infected and drug-cured people: implications for the mechanism underlying drug cure. *J Inf Dis* 184: 112

177. RJ North, 1985. Down regulation of the antitumor response. *Adv Cancer Res* 45: 1

178. Beyer M, and JL Schultze, 2006. Regulatory T cells in cancer. *Blood* 108: 804

179. GP Dunn, LJ Old and RD Schreibner, 2004. The thee Es of cancer immunoediting. *Ann Rev Immunol* 22: 329

180. W Zou, 2006. Regulatory T cells, tumor immunity and immunotherapy. *Nature Reviews Immunol.* 6: 295

181. T Yamaguchi and S Sakaguchi, 2006. Regulatory T cells in immune surveillance and treatment of cancer. *Sem Canc Biol* 16: 115

182. D Hamilton, N Ismail, D Kroeger, C Rudulier & P Bretscher, 2009. Macroimmunology and Immunotherapy of Cancer. *Immunother* 1: 367.

183. RJ North, 1986. Radiation-induced, immunologically mediated regression of an established tumor as an example of successful therapeutic immunomanipulation. *J Exp Med* 164: 1652

184. S Onizuka, I Tawara, J Shimizu, S Sakaguchi, S., T Fujita, and E Nakayama, 1999. Tumor rejection by in vivo administration of anti-CD25 (interleukin-2 receptor alpha) monoclonal antibody. *Canc Res* 59: 3128

185. DH Hamilton and PA Bretscher, 2008. Different immune correlates associated with tumor progression and regression: implications for prevention and treatment of cancer. *Cancer Immunol Immunother* 57: 1125

186. M Larche and DC Wraithe, 2005. Peptide-based therapeutic vaccines for allergic and autoimmune diseases. *Nature Med* 11: S69

187. SM Lierberman and TP DiLorenzo, 2003. A comprehensive guide to antibody and T-cell responses in type 1 diabetes. *Tissue Antigens* 62: 259

188. K Haskins, 2005. Pathogenic T-cell clones in autoimmune diabetes: more lessons from the NOD mouse. *Adv Immunol* 87: 123

189. AD Higgins, MA Milhalyo and AJ Adler, 2002. Effector CD4 cells are tolerised upon exposure to parenchymal self antigen. *J Immunol* 169: 3622

190. M Nasreen, TM Waldie, CM Dixon and RJ Steptoe, 2010. Steady-state antigen-expressing dendritic cells terminate CD4+ memory T-cell responses. *Eur J Immunol* 40: 2016

191. A Nadim A, E Javadian, G Tahvildar-Bidruni and M Ghorbani, 1983. Effectiveness of leishmanization in the control of cutaneous leishmaniasis. *Bull Soc Pathol Exot Filiales.* 76: 377

192. KB Cease and JA Berzovsky, 1994. Towards a vaccine for AIDS. *Ann Rev Immunol* 12: 923

193. LQ Zhang, P MacKenzie, A Cleland, EC Holmes, AJ Brown and P Simmonds, 1993. Selection for specific sequences in the external envelope protein of HIV-1 upon primary infection. *J Virol* 67: 3345

194. J Salk, PA Bretscher, PL Salk, M Clerici and GM Shearer, 1993. A strategy for prophylatic vaccination against HIV. *Science* 260: 1269

195. M Clerici and GM Shearer, 1993. A TH1-->TH2 switch is a critical step in the etiology of HIV infection. *Immunol Today* 14: 107

196. C Power, TW Marfleet, LF Qualtiere, W Xiao and P Bretscher, 2010. Development of Th1 imprints to rBCG expressing a foreign protein: Implications for immunization against HIV-1 and diverse influenza strains. *J Biotech Biomed* vol 2010

197. J Cohen (2001). Shots in the Dark: The Wayword Search for an Aids Vaccine. New York, WW Norton

198. N Ngo-Giang-Huong, D Candotti, A Goubar, B Autran, M Maynart, D Sicard, J-P Clauvel, H Agut, D Costagliola and C Rouzioux, 2001. HIV-specific IgG$_2$ antibodies: Markers of Helper T Cell Type 1 Response and Prognostic Marker of Long-Term Nonprogression. *Aids Res and Hum Retroviruses* 17: 1435

199. S LeClerq and PA Bretscher, 1987. T cells expressing delayed type hypersensitivity can be derived from a humorally immune lymphocyte population. *Eur J Immunol* 17: 949

200. M Altfield and BD Walker, 2001. Less is more? Supervised treatment interruption in acute and chronic HIV-1 infection. *Nat Med* 7: 881

201. FO Sanchez, JI Rodriguez, G Agudelo and LF Garcia, 1994. Immune responsiveness and lympokine production in patients with tuberculosis and healthy contacts. *Infect Immun* 62: 5673

202. L Lenzini, P Rottoli and L Rottoli, 1977. The spectrum of human tuberculosis. *Clin Exp Immunol* 27: 230

203. R Koch, 1890. Remedy for tuberculosis. *Br Med J* 1193

204. R Virchow, 1891. The effect of Koch's remedy on the internal organs of tuberculous patients. *Br Med J* 127

205. PP Lee, D Zeng, AE McCauley, YE Chen, C Geiler et al. 1997. T-helper 2 dominant anti-lymphoma immune response is associated with fatal outcome. *Blood* 90: 1611

206. MS Lowes, GA Bishop, K Crotty, RS Barnetson and GM Haliday, 1997. T helper 1 cytokine mRNA is increased in spontaneously regressing primary melanomas. *J. Invest. Derm.* 108: 914

207. M Sato, S Goto, R Kaneko, S Sato and S Takeuchi, 1998. Impaired production of Th1 cytokines and increased frequency of Th2 subsets in PBMC from advanced cancer patients. *Anticancer Res.* 18: 3951

208. P Pellegrini, AM Berghella, TD Beato, S Cicia, D Adorno and CU Casciani, 1996. Disregulation in TH1 and TH2 subsets of CD4$^+$ T cells in peripheral blood of colorectal cancer patients and involvement in cancer establishment and progression. *Cancer Immunol Immunother* 42: 1

209. K Matsumoto, H Yoshikawa, T Yasugi, S Nakagawa, et al, 1999. Balance of IgG subclasses toward human papillomavirus type 16 (HPV16) L1-capsids is

a possible predictor for regression of HPV16-positive cervical intraepithelial neoplasia. *Bioch Biophys Res Comm* 258: 128

210. M Awwad and R J North, 1989. Cyclophosphamide-induced immunologically mediated regression of a cyclophosphamide-resistant murine tumor: a consequence of eliminating precursor L3T4+ suppressor T-cells. *Canc Res* 49: 1649

About the Author

Peter Bretscher is a well-known immunologist. In 1970 he published, with Melvin Cohn, the Two Signal Model of Lymphocyte Activation, which provided an explanation for how self non-self discrimination is realized.

This theory has stood the test of time and is a central component, along with the Clonal Selection Theory, of modern immunological thinking.

Peter studied physics as an undergraduate at Cambridge University, UK, and then, hoping to find a field offering opportunities for theoretical insight, undertook graduate studies in protein X-ray crystallography in the now famous Cambridge Laboratory of Molecular Biology. He became fascinated at this time by immunology, and was fortunate in being able to discuss his early ideas with Francis Crick. This was the beginning of an almost 50 year engagement, during which Peter and his students have made substantial theoretical and experimental contributions to the field.

www.ingramcontent.com/pod-product-compliance
Lightning Source LLC
Chambersburg PA
CBHW081256170526
45165CB00011B/3318

* 9 7 8 1 4 6 0 2 7 4 0 6 4 *